Micropaleontology

Pratul Kumar Saraswati • M.S. Srinivasan

Micropaleontology

Principles and Applications

Springer

Pratul Kumar Saraswati
Department of Earth Sciences
Indian Institute of Technology Bombay
Mumbai, India

M.S. Srinivasan
Department of Geology
Banaras Hindu University
Varanasi, India

ISBN 978-3-319-79199-9   ISBN 978-3-319-14574-7   (eBook)
DOI 10.1007/978-3-319-14574-7

Springer Cham Heidelberg New York Dordrecht London
© Springer International Publishing Switzerland 2016
Softcover re-print of the Hardcover 1st edition 2016

Printed on acid-free paper

Springer International Publishing AG Switzerland is part of Springer Science+Business Media
(www.springer.com)

Preface

Microfossils have been mentioned as early as the fifth century BC by Herodotus, who travelled to Egypt and noted large size foraminifer *Nummulites* in the pyramids. The history of the study of microfossils since then has seen some major turning points; the major ones being the invention of the microscope in the late seventeenth century, followed by the voyage of *HMS Challenger* (1872–1876) - the first global marine research expedition, increased activity in exploration of oil after the second world war (1939–1945), and the voyage of the *Glomar Challenger* by an international initiative to core deep-sea sediments (Deep Sea Drilling Project, 1968). The development of micropaleontology through these years has unequivocally established the multidimensional applications of microfossils in solving problems in Earth Science. The oil industry nurtured it since the twentieth century and its core application in subsurface stratigraphy continues in spite of commendable development in geophysical logging and seismic methods to interpret subsurface geology. In fact, micropaleontology has expanded its role by being indispensable in seismic calibration, sequence stratigraphy, and as a tool to guide horizontal drilling of the reservoirs to save the cost of production. The Deep Sea Drilling Project and its successors revealed another dimension of micropaleontology by providing vital information about ocean currents, deep-sea processes and paleoclimates of the past. Microfossils thus became an integral part of the newly developed field of paleoceanography. It may not be inappropriate to say that among the various tracers of earth history, microfossils provide a wealth of information including the evolution of life, age and paleoenvironment of sedimentary strata, paleoclimate and paleoceanography.

The multiple applications of microfossils attract students and professionals of various backgrounds to employ microfossils in their research. The subject is usually taught at the postgraduate levels of universities, and it is expected that the student has a basic understanding of the principles of paleontology. Most textbooks generally focus on principles of paleontology in reference to mega invertebrate fossils. However, several issues regarding sample collection, taphonomy, biomineralization and ecological details that are unique to microfossils also need to be explored and described. Keeping this in mind, Part I of the textbook addresses the basic principles of micropaleontology with examples of microfossils so that those who don't have a formal background in paleontology can also develop an understanding of the systematic approach to the study of microfossils. It is important to be aware of the strengths and limitations of microfossils and their geological records. Part II gives an overview of the major groups of microfossils including their morphology, ecology and geologic history. Marine microfossils, particularly foraminifera, are discussed in greater detail compared to other groups as they continue to play a major role in most scientific investigations. Part III of the book explains the applications of microfossils in biostratigraphy, paleoenvironmental interpretation and paleoclimate reconstruction, basin analysis for hydrocarbon exploration and paleoceanography.

We gratefully acknowledge and thank our colleagues and students who have contributed in various ways in preparing the book: Santanu Banerjee, Anupam Ghosh, Sonal Khanolkar, Asmita Singh, Arundeo Singh, Jyoti Sharma, Ajai Rai, C. N. Ravindran, Jahnavi Punekar, Johann Hohenegger and Komal Verma. Thanks are due to Pradeep Sawant who drafted all the illustrations. We thank Sherestha Saini, Editor, Springer Science + Business Media, for initiating the proposal and bringing it to its conclusion. P.K.S. owes more than formal thanks to Indian Institute of Technology, Bombay for providing the academic freedom to work and granting a sabbatical to write the book. M.S.S. expresses his gratitude to Banaras Hindu University for enabling him to carry out continued research in micropaleontology.

P.K. Saraswati
M.S. Srinivasan

Contents

Part III   Applications

Part I
Principles

Chapter 1
Introduction

1.1 Introduction

Microfossils attracted naturalists long before the invention of the microscope by van Leeuwenhoek (1632–1723). The Greek philosopher Herodotus was possibly the first to notice microfossils in the fifth century B.C., although in a different sense. The large-sized foraminifer *Nummulites* in the Eocene limestone slabs that make up the Egyptian pyramids was thought to be petrified lentils eaten by the slaves who constructed them. Microfossils played no significant role in the early development of paleontology from the sixteenth to the nineteenth century, but it has had phenomenal growth since the early part of the twentieth century. The growth of micropaleontology is sustained due to wide-ranging applications of microfossils in solving diverse geological problems. Oil professionals searching for marker horizons and stratigraphic tops in oil fields stimulated its early success. The initiation of deep-sea exploration in the 1960s led to the expansion of micropaleontology in the areas of paleoceanography and paleoclimatology. It provided an unprecedented opportunity to study the evolutionary processes and phenotypic variation within various groups of microfossils, many of which had hardly been studied up until then. The use of microfossils in applied geological investigations is becoming more and more important. Today, micropaleontologists have begun addressing contemporary issues of environment and climate change using microorganisms, the fossilized remains of which have so far been the principal objects of their study.

Micropaleontology is the systematic study of microfossils, their morphology, classification and environmental and stratigraphic significance. For practical purposes, a microfossil is any fossil, usually small, whose distinguishing characteristics are best studied by means of a microscope. It includes a heterogeneous group of fossils of organisms that are generally of microscopic size, for example, foraminifera, ostracoda and radiolaria. It also includes the detached skeletal elements of large-sized organisms that are difficult to identify without the aid of

© Springer International Publishing Switzerland 2016
P.K. Saraswati, M.S. Srinivasan, *Micropaleontology*,
DOI 10.1007/978-3-319-14574-7_1

Table 1.1 Types of microfossils based on test composition, their size range, environmental distribution and geologic age

Microfossil	Test composition	Size range	Environment	Geologic age
Foraminifera	Calcareous, agglutinated, siliceous, chitinous	0.1 mm–10 cm	Marine	Cambrian–recent
Ostracoda	Calcareous, chitinous	0.5–5 mm	Marine–freshwater	Ordovician–recent
Calcareous nannoplankton	Calcareous	<20 μm	Marine	Jurassic–recent
Pteropods	Calcareous	2.5–10 mm	Marine	Eocene–recent
Radiolaria	Siliceous, strontium-sulphate	100–2000 μm	Marine	Cambrian–recent
Silicoflagellates	Siliceous	20–100 μm	Marine	Cretaceous–recent
Diatoms	Siliceous	5–2000 μm	Marine–freshwater	Cretaceous–recent
Conodonts	Phosphatic	0.1–3 mm	Marine	Cambrian–Triassic
Spores/pollens	Organic	5–200 μm	Terrestrial, brackish to freshwater	Silurian–recent
Dinoflagellates	Organic	20–150 μm	Marine–freshwater	Permian–recent
Acritarchs	Organic	<100 μm	Marine	Precambrian–recent
Chitinozoa	Organic	30–1500 μm	Marine	Ordovician–Devonian

a microscope, such as sponge spicules, plates of corals and echinoderm spines. Notwithstanding their heterogeneity, the major groups of microfossils are classified based on their shell composition (Table 1.1). It may be noted that such a grouping does not suggest their biological affinities. What binds them as a subject matter of micropaleontology is their minute size, abundant occurrence, the special techniques for their study and their outstanding contributions to different aspects of geology.

1.2   Major Groups of Microfossils

All the six kingdoms of life, archaebacteria, eubacteria, protista, animalia, plantae and fungi, have microscopic life forms. Only a few of them, however, are commonly applied in the interpretation of ancient environments and in the dating and correlation of sedimentary successions. Some of the geologically useful microfossils are illustrated in Fig. 1.1 and described below in brief.

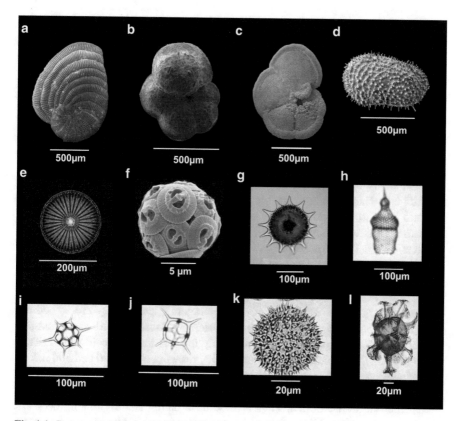

Fig. 1.1 Representative microfossils belonging to Foraminifera (**a, b, c**), Ostracoda (**d**), Diatoms (**e**), Calcareous nannoplankton (**f**), Radiolaria (**g, h**), Silicoflagellate (**i, j**), Pollen (**k**) and Dinoflagellate (**l**)

Foraminifera: These are single-celled animals belonging to the phylum Protista. Nearly 50,000 species of this group have been described, of which over 4000 species are living today. The shells, called test, are mostly calcareous, though a few of them are pseudo-chitinous, agglutinated or, rarely, siliceous. Their size ranges from about 0.1 mm to more than 10 cm. Architecturally, the tests are of diverse types, and due to distinct morphological characters, the species are readily identifiable. Foraminifera are of both types, benthic and planktic. The benthic species are useful in paleoenvironmental interpretation and the planktic species have been successfully used in biostratigraphy, particularly for the Cretaceous and younger sequences. Stratigraphically, they occur in rocks of all ages from the Cambrian times onwards. In certain parts, they were so abundant that they formed huge thicknesses of limestone, some of the most famous being the *Fusulina* limestone of the Permo-carboniferous age and the Nummulitic limestone of the Eocene age. Today, they occupy practically all the ecological niches from marginal marine to deep marine environments. Most are marine and stenohaline, but certain groups can occur under hypersaline conditions. Some species prefer water with low salinity, and they frequently occur in brackish lagoons and estuaries.

Calcareous Nannoplankton: The calcareous nannoplankton are unicellular, autotrophic marine algae called coccolithophores. Their size can range up to 60 μm, though they are usually less than 20 μm. Their established first occurrence was in Jurassic sediments, though they may extend into the Triassic or even Paleozoic. They inhabit waters well lit for photosynthesis, poor in nutrients and rich in oxygen. Their maximum abundance, found in the tropics, is, at most, 50 m. Most of the species prefer warm and temperate waters. They are largely adapted to normal salinity, and therefore occur in the open ocean, but some species have been found both in coastal and fresh waters.

Ostracoda: These are crustaceans, ranging in size from 0.5 to more than 5 mm. Their shells, called carapace, are made of two valves and may be calcareous or chitinous. They have both benthic and planktic modes of life. Their fossil record dates back to the Ordovician. The ostracoda live in all aquatic environments, from oceans to estuaries, lagoons, freshwater ponds and even in moist soils. Most of the Paleozoic ostracoda were marine. The first lacustrine ostracoda appeared in the late Carboniferous. The post-Paleozoic ostracoda achieved the same level of ecological diversity as seen today.

Pteropods: These are marine, planktic gastropods. Though not very common in the fossil record due to solution-susceptible aragonite shells, they are sometimes found in abundance in Neogene and Quaternary sediments. Most present-day pteropods live in tropical waters, with only few occurring in cold waters. They occur at all depths but are most abundantly present between 300 and 500 m.

Radiolaria: The radiolarians are marine protists made of siliceous skeletons beautifully formed from rods and lattices. They are marine and planktic and first appeared in the Cambrian. They are useful in dating and correlating deep-sea sedimentary rocks. All the living species are marine and stenohaline, preferring a salinity of >30‰. Their maximum abundance is at a depth of 100 m and they decline towards the surface waters and below depths of 500 m. The environmental distributions of fossil radiolarians of the Cenozoic and Mesozoic were largely similar to that of today's radiolarians. Some radiolarians associated with corals and rudist bivalves suggest that they inhabited shallow marine environments close to continents.

Diatoms: The diatoms are algal forms, and differ from other algae in having siliceous cell walls. Their skeletons, called frustules, occur both in marine and freshwater sediments of the Jurassic to recent times. At times, they have formed soft rocks, called diatomite, constituted chiefly of diatom frustules. Many diatoms occur in extreme temperature conditions, from polar ice caps to hot thermal springs.

Silicoflagellates: Silicoflagellates are marine phytoplanktons, recorded since Cretaceous times. These unicellular chrysophyte algae have discoidal or hemispherical skeletons of opaline silica. The silicoflagellates are useful in the correlation of deep-sea sediments. In modern oceans, they tolerate salinity variation of 20–40‰ and occur in all temperature regimes, from polar to tropical.

Dinoflagellates: These are unicellular algae, occurring as parasites, symbionts or as independently living autotrophic plankton of the surface waters of marine, lagoonal and lacustrine environments. The cysts of dinoflagellates have been recorded in Silurian, Permian and Triassic strata; however, they become common only in Jurassic and younger sediments.

Acritarchs: These are organic-walled microfossils, similar to dinoflagellate cysts, but with uncertain biological affinity. A majority of acritarchs are unicellular phytoplanktons and marine, but some of the Holocene species have been recorded in fresh water. Acritarchs evolved in the Precambrian (~3.2 billion years ago) and reached their acme in the Ordovician. Acritarchs are useful in deciphering the maturity of hydrocarbon source rocks.

Chitinozoa: These are organic-walled microfossils of uncertain affinity. The flask-shaped vesicles of chitinozoans range mostly from 150 to 300 μm long. They first appeared in the Early Ordovician and became extinct in the Devonian. Chitinozoans were exclusively marine and inhabited settings from the shelf to the basin. They are a good indicator of the thermal history of sedimentary basins.

Spores and Pollens: These are reproductive organs of higher plants. Though not mineralized, their organic constituents are durable and preserved in sediments. Due to chemical and mineralogical modifications during burial, there is progressive change in the colour of spores and pollens from pale yellow to black through orange and brown. This is used as an indicator of the temperature regime of organic maturation in sedimentary basins. The earliest known spores are from the Silurian and the pollens appear in the Devonian and diversify in the Carboniferous. The angiosperm pollens became abundant in the Cretaceous. The distribution of spores and pollens depends on the ecology of their parent plants, but, depending upon the size, weight and atmospheric conditions, the spores and pollens may disperse in all environments, from terrestrial to fresh, brackish and marine waters.

Conodonts: The phosphatic microfossils belonging to conodonts are a widespread and biostratigraphically important group of microfossils for the Paleozoic. The alteration of conodont elements, expressed as the "conodont colour alteration index", is an indicator of thermal maturity, and hence useful in basin analysis for hydrocarbon exploration. The biological affinity of conodonts was debated for a long time, but well-preserved fossils indicate that they are an extinct group of primitive jawless vertebrates belonging to the Chordata. They occur as tooth-like isolated hard parts in sediments of the Cambrian to Triassic ages.

1.3   Collection of Samples

Outcrop Samples

Only large-sized microfossils such as larger benthic foraminifera can be collected in the field in the same way as megafossils. Otherwise, collectors of microfossils see nothing of the fossils collected until they are recovered and separated from the matrix in the laboratory. They do not collect fossils in the field, but samples of rock that they believe may yield microfossils on maceration. Experience is, therefore, essential in collecting samples for micropaleontological investigations. Because of their small size and fragile nature, microfossils are carefully collected and preserved. Some rock types have a higher potential of yielding microfossils than others. Limestone, dolomite, calcareous mudstone and thin parting of claystone in fossiliferous limestone are usually rich in microfossils. Many carbonate rocks are exceptionally fossiliferous, and in some instances, may be made up primarily of foraminifera and other microfossils. Some clastic rocks, such as fine-grained sandstones, siltstones and glauconitic sandstones, at times contain a good number of microfossils. The coarser clastics, as a rule, are poor in well-preserved forms. Black shales, chalk and cherts often contain well-preserved microfossils. The soft and easily washed rocks are especially scrutinized because of easy disaggregation for separation of microfossils.

The micropaleontological sampling should be carefully planned. It is common to observe microfossils being concentrated in some thin bands in an otherwise thick monotonous succession. Such horizons may easily be missed if proper sampling procedures are not followed. Spot sampling at regular intervals is best employed for thick sections of essentially uniform lithology and for thin shale or clay beds in sandstone or limestone. Channel samples are taken in short sections of, say, a few tens of metres. A limitation with channel sampling is that the exact stratigraphic position is blurred, but its advantage is that the fossiliferous horizons are not missed. It is a standard procedure to collect samples from strata containing megafossils.

The samples are collected at regular intervals that depend upon the objective of the work, the time available and the thickness and character of the strata. The sample locations are marked on the toposheet and GPS readings are noted. Graphic logs are prepared for the traversed section and stratigraphic positions of the collected samples are marked in them. The usual samples taken for micropaleontological analysis are 200–250 g. The collector of microfossils should carry a Brunton or clinometer compass, a hammer, a chisel, tape, a hand lens, thick plastic and cotton bags, an acid bottle containing 10 % solution of HCl, and a paint brush for clearing rock surfaces. The most important precaution to be exercised during sample collection is to avoid contamination. The hammers and chisels should be cleaned before each sample is collected. Sample bags should be labeled on the spot and details entered in a field notebook. Non-labeled or poorly labeled samples are of no scientific value.

Subsurface Samples

The subsurface samples are mainly obtained by rotary drilling. This involves rotating a drill bit attached to a hollow steel tube (drill string). A commonly used drill bit consists of three rotating cones set with teeth. The teeth chip away the rock at the borehole bottom as the bit is rotated. The drill cuttings are removed from the borehole by the drilling mud. The micropaleontological studies based on cutting samples should be interpreted cautiously, due to the possibility of

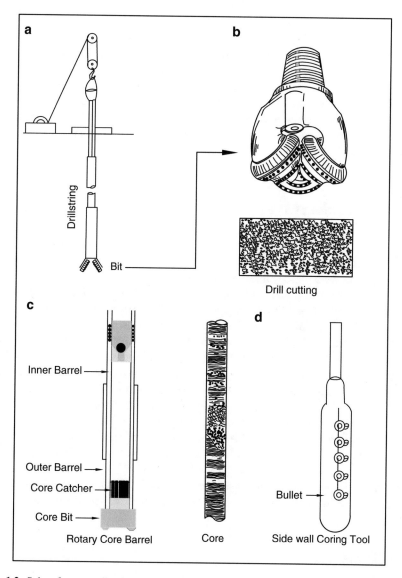

Fig. 1.2 Subsurface sampling by drilling: The rotating system (**a**) comprises a drill string with a drilling bit (**b**) attached at the end for drill cutting of the formation or a rotary core barrel (**c**) to obtain core samples. A sidewall-coring tool (**d**) is lowered into the borehole to get sidewall core samples

contamination by caved materials from the overlying formations. Another type of subsurface sample is recovered by coring. The core sample is recovered by a core barrel, which is a hollow steel tube fitted with industrial diamonds or other grinding agent at the down-hole end. As the core barrel is rotated, it cuts a cylinder of rock that is retrieved when the core barrel is withdrawn back to the surface (Fig. 1.2). Coring is slower and more expensive than drilling with ordinary bits. The core samples, because of their undisturbed stratigraphic positions, are most reliable for micropaleontologic studies. However, due to the higher cost, these are normally only recommended at crucial intervals where either reservoir properties are to be studied in detail or stratigraphic boundaries are to be precisely marked. Sidewall cores are recovered from already drilled boreholes by firing hollow cylindrical bullets into the penetrated formation. Although the cost is comparatively less, only small amounts of samples are recovered in this method. Such samples may be recommended where stratigraphic ranges of age-diagnostic taxa need to be confirmed in a section.

Deep Sea Drilling Core Samples

The Deep Sea Drilling Project (DSDP) in the late 1960s and its successor, the Ocean Drilling Program (ODP) during the mid-1980s, made available deep-sea core samples which hitherto were inaccessible. The development of two new coring devices in the 1980s provided a major boost for oceanic micropaleontology and paleoceanography. The conventional rotary drilling techniques yielded highly disturbed sedimentary cores, a problem that was solved with the Hydraulic Piston Corer (HPC) since it rapidly punched into sediment without rotation (Fig. 1.3). The hydraulic piston corer is used in the most unconsolidated upper part of the section, and sequences of at least up to 200 m can be cored in this manner. However, when sediments become too compacted to allow use of the hydraulic piston corer, the Extended Core Barrel (XCB) is employed. The XCB provided new opportunities for high resolution stratigraphic and paleoceanographic studies over much longer intervals of time than conventional piston cores, which are very restricted in length. At present, it is possible to drill as far as 2 km under the seafloor, but this is not enough to address current research issues. Now, in Japan, the Japan Marine Science and Technology Center (JAMSTEC) has built a large research ship equipped with a riser drilling system. This research ship, called the "CHIKYU", is expected to revolutionize marine science research in the twenty-first century.

Recent Samples

The recent sediments may be collected by a grab or a gravity corer. The live specimens of foraminifera and other benthic marine organisms can also be collected by scuba diving. The planktic organisms are collected from a water column by a plankton net with a mesh opening of generally 10–35 μ. The living organisms may be preserved in solutions of 10 % ethanol or buffered formalin.

It is important in biologically oriented studies to distinguish live specimens in the total assemblage. There are a number of techniques for determining if foraminifera are alive at the time of collection. The techniques are broadly categorized as terminal and non-terminal (Bernhard 2000). The terminal methods

Fig. 1.3 Piston corer (*a*)
to recover deep-sea core
samples and a grab
sampler (*b*) for collecting
surface sediments

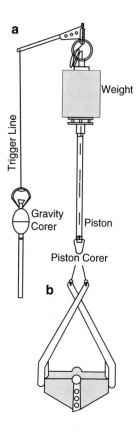

kill the specimens to know if they were alive, and include staining, Adenosine-5′-triphosphate (ATP) assay and ultrastructural studies of foraminiferal cytoplasm by transmission electron microscopy. The non-terminal techniques do not kill the foraminifera. These include cytoplasm colour, accumulation of debris around the apertural region, observation of the pseudopodial network under a phase contrast microscope and fluorescence spectroscopy. The live specimens of larger benthic foraminifera are readily identified in the field by the greenish colour of the cytoplasm imparted by the symbionts.

In spite of certain limitations, the staining method is the most popular for differentiating live and dead forms. The most commonly used stains are Rose Bengal and Sudan Black B. To stain the samples, the washed residue on a 240 mesh screen is placed in a bowl of 1 g/l solution of Rose Bengal stain. The sample is allowed to stain for 20 min. The residue is then washed under a fine spray of water to remove stains from the sediment grains and shell walls. The protoplasm of foraminifera is stained deep red, and thus can be examined under the microscope. The shells with opaque walls may have to be broken to see the red coloration. The efficacy of the staining technique has been questioned by some workers who suggest the combined use of Adenosine-5′-triphosphate (ATP) assay and vital staining technique for more accurate estimation of living populations.

A standardized protocol for sampling of modern benthic foraminifera is suggested based on the recommendations of the Foraminiferal Biomonitoring (FOBIMO) workshop (Schonfeld et al. 2012). The mandatory recommendations are as follows: (1) The sample should cover the interval 0–1 cm below the surface; (2) An interface corer or box corer should be used for offshore surveys so that the sediment surface is kept intact; (3) Three replicate samples should be taken and analysed separately; (4) Samples are to be washed on a 63 μm screen and the living benthic foraminifera of >125 μm fraction is to be analysed; (5) Splits are to be picked and counted entirely; (6) For staining the Rose Bengal with a concentration of 2 g/l, and staining time of at least 14 days should be used.

Molecular taxonomic studies are gaining importance in micropaleontology. Presently, much of the work is concerned with DNA extraction of the living representatives, especially of foraminifera. For DNA extraction, it should be ensured that the specimens were alive at the time of collection. Staining is not a sufficient criterion for distinguishing live specimens. In foraminiferal samples, the best way is to pick the individuals showing extended pseudopodia. The selected samples may be stored by freezing at –20 °C. Air-dried samples stored at room temperature can also be used if they have not been left for a longer time. The samples intended for molecular analysis should not be preserved in ethanol or formaldehyde (refer to Holzmann and Pawlowski 1996 for further details).

1.4   Separation of Microfossils from Matrix

The separation of microfossils from the matrix involves a combination of mechanical and chemical disaggregation techniques. The exact technique depends on the chemical composition of the shell wall and the nature and age of the sediments hosting the microfossils. The quality of recovery is improved by skill, experience and experimentation. In general, the less treatment necessary to process the sample, the less damage to the microfossils. The minimum amount of crushing, boiling and sieving should be used in order to ensure minimum breakage of specimens. Some general techniques applicable to major groups of microfossils are discussed below.

Calcareous Microfossils

For extraction of foraminifera and ostracoda, the samples are coarsely broken up into fragments of about 1 cm in size. It is a common practice to divide the samples into two equal parts, one of which is preserved for reference or use in case unsatisfactory results are obtained from the other half-sample.

Softer and poorly consolidated samples, such as calcareous shale and mudstone, may be disaggregated by simply soaking them in water for a few hours. The disaggregated material is then washed through a sieve with openings of 63 μm (ASTM 230 mesh). This sieve size retains a majority of the microfossils and their juveniles. Wet screening is very satisfactory in most cases. Sometimes, a gentle rubbing of small lumps by hand is advisable, but with utmost care. The washed residue may then be dried on a hot plate or in an oven. The dried residue is placed on a picking tray for inspection under a stereoscopic binocular microscope. Moderately hard samples may not yield by a simple soaking. Such samples may be disintegrated by boiling the mixture of sample and water with sodium bicarbonate, sodium hexametaphosphate or sodium hydroxide, all of which are deflocculating agents. The boiling should continue for an hour or so until the material is disaggregated. The sample is then decanted, washed and dried. An alternative method for processing hard mudstones is to soak the sample overnight in 50 % H_2O_2 and boil the solution the next day over low heat for half an hour or so. A little water is added afterward and the mixture is boiled again. The sample is then washed and dried. A chemical of trade name Quat-O also satisfactorily disaggregates moderately hard samples.

For harder rocks, such as some argillaceous sandstone shales and marls, other methods of processing have to be followed. Excellent results have been obtained with the crystallization of photographic hypo (sodium thiosulphate). The dry sample is broken into smaller pieces and covered with a concentrated solution of sodium thiosulphate or mixed with an equal amount of crystals and then heated. When soaked with the solution, the sample is kept in a cold place, where it remains until the salt crystallizes and forms a solid mass together with the rock fragments. The sample is then melted. This process is repeated several times until most of the rock is broken up by the growth of crystals in pores and fissures. The disintegrated rock sample is then washed carefully until the salt is completely removed. Removal of microfossils from chert and other siliceous sediments is difficult. Many such samples are best studied in thin sections or by immersing thin chips of the material in alcohol or immersion oil.

Precaution is necessary at all stages of sample preparation. The equipment must be kept thoroughly clean to prevent contamination of one sample by others, and it is a good practice to record each sample in a register as it is washed, so that contamination, if any, can subsequently be traced. When sieves are used, a simple check on contamination can be made by dipping them in a solution of methylene blue after washing each sample. Blue specimens in washed concentrates are then rejected.

In the final stage, microfossils from the dried residue (comprised of mineral grains, fine rock fragments and microfossils) are separated. Separation of the fossil remains from the residue may be done by use of heavy liquids of varying densities. Carbon tetrachloride or bromoform is used to float and concentrate foraminifera from the residue. These chemicals are toxic and, therefore, should be handled with care under a fume hood. Sodium polytungstate is a safer agent for flotation, as it is non-toxic. The most common practice for removing microfossils is to pick individual specimens from the various screened fractions of the residue with a fine camel's hair brush under a stereozoom binocular microscope and mount them on a gummed micropaleontological slide. The modest requirement for picking of microfossils includes a picking tray (the inner surfaces of which are painted black and its

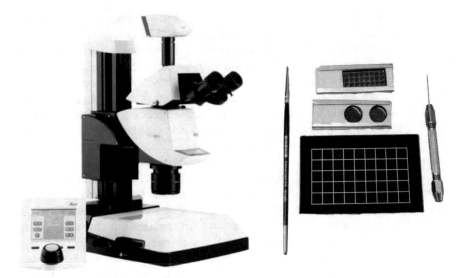

Fig. 1.4 Stereozoom binocular microscope, picking tray, brush, needles and microfaunal slides for picking and storing microfossils

grids marked), a fine brush of 00 or 000 size, dissecting needles, forceps, water-soluble adhesive (such as gum tragacanth) and micropaleontological slides (Fig. 1.4). Skill in using the moistened and pointed brush for picking up the specimens under the microscope can be acquired only with practice.

Many foraminifera, particularly the larger ones, can be identified only by their internal structures. This requires preparing oriented thin sections to examine chambers, wall structures, marginal canals, tooth plates and other features. Equatorial and axial sections are required for observation of internal features. The procedure for making sections is as follows:

A mounting medium, Canada balsam or lakeside 70c thermoplastic, placed on a glass slide, is gently heated over a spirit lamp or on a hot plate until it melts. The dry specimen is immersed in the melt, the slide is transferred to a microscope while still warm, and the specimen is positioned with a heated mounting needle. The specimen is held in place by the mountant as it cools. The specimen mounted on the glass slide is then rubbed on a glass plate by using successively finer carborundum powder and clean water is added constantly for lubrication. The grinding is monitored continuously under a microscope until the desired plane is reached. The section is then washed thoroughly to remove the grinding powder. The slide is heated again to melt the mountant and the specimen is turned over with a hot mounting needle so that the ground side is against the glass. After cooling, the grinding is resumed from the other side of the specimen until the section is sufficiently thin. It is not necessary to make ultra-thin sections. The section is cleaned of the grinding powder and a coverslip is mounted. Sometimes, when the chambers are not filled up by secondary calcite, a split section along the equatorial plane can be obtained by heating the specimen in a flame and quenching it in ice-cold water.

Calcareous Nannoplankton

The calcareous nannoplankton can be extracted from fine-grained marine sediments by centrifuging approximately 5 g of crushed rock sample that are passed through a 30 mesh sieve. Distilled water is added and the mixture is thoroughly stirred. The sample is short centrifuged at 300 rpm for 15 s. The decantant is set aside and the process is repeated 3–8 times until the decantant is nearly clear. The decantant is short centrifuged again at 850 rpm for 30 s and then set aside for later examination. The residue contains particle size between 3 and 25 μm, rich in nannofossils. To prepare the strewn slides, the suspension containing nanoplankton is dispersed on a glass coverslip and dried. The coverslip is mounted on a glass slide with Canada balsam. For scanning electron microscopic observation, the solution may either be directly dispersed on the stub or dispersed on a coverslip and dried. The slides are examined under a high-powered petrological microscope with magnification of 1000–1500× and oil-immersion lenses.

Siliceous Microfossils

The processing techniques for extracting siliceous microfossils, such as radiolaria, diatoms and silicoflagellates, essentially involve treatment with hydrogen peroxide to remove organic matter and hydrochloric acid to remove carbonate fractions. After treating the sample with HCl, the suspension is decanted and the residue is further treated with 25 % HNO_3 to remove the last traces of organic matter and inorganic salts. Distilled water is added to the residue and allowed to stand for sometime. The solution is decanted and the residue is ready for preparation of the slide. To prepare the slide, the residue is diluted with distilled water and stirred well so that the microfossils are brought into suspension. A small amount of this suspension is spread on a glass slide and dried on a hot plate. It can be fixed in glycerine for temporary mounting or in Canada balsam or hyrax for permanent mount.

Phosphatic Microfossils

For extraction of conodonts, 1–2 kg of rock samples is required for acid treatment. The crushed rock samples are immersed in 10 % glacial acetic acid (CH_3COOH) buffered with sodium acetate or calcium acetate to a pH of 3.5. Alternatively, 10 % formic acid (HCOOH) is buffered with calcium carbonate and tricalcium carbonate to a pH of less than 3.6. The digestion may take a day to several days and require regular changing of the acid.

Plant Microfossils

The preparation procedures for plant microfossils consist of dissolution of rocks by a succession of chemical reagents. The crushed rock fragments, placed in a polypropylene beaker, are first reacted with HCl and then with HF to remove the carbonate and silicate fractions of the rock. The residue is washed, centrifuged and then reacted with Schulze solution, which is prepared by mixing one part saturated aqueous solution of $KClO_3$ with two parts concentrated HNO_3. The residue is washed and treated with 10 % solution of KOH or NaOH to remove humic materials. The final residue consists of spores, pollens, microplankton, leaf and bark cuticles, and other plant fragments. It is centrifuged, dried and mounted on slides with coverslips for examination under a light microscope. Extreme care should be taken in processing palynological samples. All acid treatments should be carried out in fume chambers, with gloves and goggles. Contamination from other samples, as well as from the atmosphere, must be avoided.

1.5 Preparation of Specimens for Scanning Electron Microscopy

The Scanning Electron Microscope (SEM) is of immense value in the study of microfossils. Due to high resolution, the surface ultrastructures of microfossils, including many extremely small types, are clearly visible under a SEM. The specimens for electron microscopic observation should be free from dirt. Before mounting on a stub, they should be placed on a glass slide and cleaned with a wetted brush or centrifuged. The cleaned specimens are mounted onto stubs with double-sided adhesive tape. An alternative method is to stick a glass coverslip to a stub with a heat-proof adhesive and mount the specimen on the coverslip with water-soluble glue, such as gum tragacanth. Mounting specimens on exposed negative film glued to a stub makes for a good background. An advantage of this procedure is that specimens can be fixed to substratum with a little moisture and can be remounted with ease for photographing in different orientations. A thin film of Au, Au–Pd, C or another suitable electrically conductive material coats the mounted specimens under a sputter coater. It is also recommended to use silver or carbon conductive paint to maintain electrical conductivity between the specimen and the stub. The environmental SEM does not require coating of specimens; the clean specimens are directly examined under the microscope.

1.6 Sample Preparation for Shell Geochemistry

The trace element composition and oxygen and carbon isotopic ratios in the calcareous shells of microfossils, mainly foraminifera, provide important clues for paleoceanographic and paleoclimatic interpretations. The low values of trace elements make it necessary that the shells be free of any extraneous contamination. Likewise, the original oxygen and carbon isotopic ratios ($\delta^{18}O$, $\delta^{13}C$) of the shells can alter due to diagenesis. In view of this, rigorous procedures are followed in cleaning the microfossil shells and in obtaining pristine shells for geochemical analysis.

All the cleaning procedures for trace element analysis of foraminifera begin with crushing about 2 mg of shells between glass plates and ultrasonicating them several times under distilled water and methanol rinses. The sample is then dissolved overnight in ammonium acetate/acetic acid buffer (pH=5.5), centrifuged, and the supernatant is removed for analysis. Better results in removing contaminants are obtained by cleaning with basic hydroxylamine hydrochloride solution or basic sodium dithionite complexing reagent at 80 °C for 30 min. The detailed protocols for cleaning and dissolution to improve trace element data are continuously being experimented with and users should refer to the latest literature in this area (e.g. Barker et al. 2003). The instruments for trace element analysis include the Atomic Absorption Spectrophotometer (AAS), the Inductively Coupled Plasma Atomic Emission Spectrophotometer (ICP-AES) and the Inductively Coupled Plasma Mass Spectrometer (ICP-MS).

For stable isotopic analysis of microfossil shells, the foremost requirement is that shells are pristine and the original isotopic composition is preserved. Samples affected by diagenesis should be eliminated. The first step should be to examine the shells under an optical microscope to see that they are not iron-stained, do not have in-fills of secondary minerals, and have clear primary wall mineralogy. The optically pristine specimens are then examined under a SEM to ensure that microstructures are clearly preserved and there is no overgrowth of secondary minerals. Samples can also be examined under a cathodoluminescence microscope to check for signs of alteration.

References

Barker S, Greaves M, Elderfield H (2003) A study of cleaning procedures used for foraminiferal Mg/Ca paleothermometry. Geochem Geophys Geosyst 4:8407. doi:10.1029/2003GC000559

Bernhard JM (2000) Distinguishing live from dead foraminifera: methods review and proper applications. Micropaleontology 46(Suppl 1):38–46

Holzmann M, Pawlowski J (1996) Preservation of foraminifera for DNA extraction and PCR amplification. J Foraminifer Res 26(3):264–267

Schonfeld J, Alve E, Geslin E, Jorissen F, Korsun S, Spezzaferri S, Members of the FOBIMO group (2012) The FOBIMO (Foraminiferal Bio-Monitoring) initiative – Towards a standardised protocol for soft-bottom benthic foraminiferal monitoring studies. Mar Micropaleontol 94–95:1–13

Further Reading

Kummel B, Raup D (eds) (1965) Handbook of paleontological techniques. WH Freeman, San Francisco

Lipps JH (1981) What, if anything is micropaleontology? Paleobiology 7(2):167–199

Chapter 2
Taphonomy and Quality of the Fossil Record

2.1 Introduction

The reliability of microfossils as stratigraphic and environmental indicators depends on how accurately the microfossil assemblage corresponds to the live community. The physical, chemical and biological factors modify the remains of organisms to varying degrees during the process of fossilization. The ability to distinguish between the in situ and transported microfossils is important in this regard. The elemental and isotopic compositions of the shells are widely used as proxies of the paleoenvironment, but is the original composition preserved in the fossil shells? The organisms undergo a complex transition from biosphere to lithosphere after death. The decay of the organic matter, post-mortem transport of the shells and their dissolution and diagenesis prior to and after burial under sediments constitute the science of taphonomy. The fossil assemblages are categorized as autochthonous, para-autochthonous and allochthonous. The autochthonous assemblage represents a community of organisms preserved in their life positions. The autochthonous components moved from their positions but not transported to another community form a para-autochthonous assemblage. The fossil assemblage derived from different communities is called allochthonous. In general, in an oceanic realm, the remains deposited on the ocean floor (thanatocoenosis) are the counterpart of distributions in the water (biocoenosis). Transport and differential dissolution during sinking of the assemblages (sidocoenosis) may, however, blur this process, resulting in assemblages on the sea floor that are better referred to as taphocoenosis.

Improved levels of biostratigraphy and dating techniques have enabled interpretations of stratigraphic records at the sub-million year scale. This, indeed, led to an understanding of short-term geologic and climatic processes and their global correlations. A cautious approach is, however, required at this resolution, particularly in correlating the interpreted processes on a global scale. There is growing practice

© Springer International Publishing Switzerland 2016
P.K. Saraswati, M.S. Srinivasan, *Micropaleontology*,
DOI 10.1007/978-3-319-14574-7_2

of extracting environmental and climatic signals at increasingly higher time resolutions to match them with observations at the scale of a human lifetime. Monsoon reconstruction from geological records, for example, is attempted at centennial and decadal scales. How practical is it to attain temporal resolution of this order? The rate and continuity of sedimentation and the time averaging of fossil assemblages determine the temporal resolutions of geologic records. Microfossils are key to wide-ranging interpretations of biological and environmental processes in the geologic past. Moreover, key to reliable interpretation is a diligent assessment of the quality of the fossil record.

2.2 Decay of Organic Matter

In the rigorous process of fossilization, the soft part of the organism is preserved only in exceptional circumstances. The state of preservation of carbonate shells is governed by the chemical reactions involved in the decay of organic matter (see Box 2.1 for chemical reactions). The organic matter gets oxidized by aerobic microbes soon after the death of the organism. The microbial decay is initially under an oxygen-rich (aerobic) environment and, in the process, nitric and phosphoric acids are produced. Once oxygen is consumed, further decay occurs under an anaerobic condition, nitrate being the source of oxygen for oxidation of organic matter. After nitrate reduction, the decay of organic matter in marine sediments occurs through sulphate reduction. There is formation of acids in the decay reactions. The

Box 2.1: Chemical Reactions of Decay of Organic Matter

Aerobic Decay:

$$(CH_2O)_{106}(NH_3)_{16}(H_3PO_4) + 138O_2 \rightarrow 106CO_2 + 122H_2O + 16NO_3^-$$
$$+ 16H^+ + H_3PO_4$$

Anaerobic Decay:

$$(CH_2O)_{106}(NH_3)_{16}(H_3PO_4) + 84.8HNO_3 \rightarrow 106CO_2 + 42.4N_2 + 148.4H_2O$$
$$+ 16NH_3 + H_3PO_4$$
$$(CH_2O)_{106}(NH_3)_{16}(H_3PO_4) + SO_4^{2-} \rightarrow 106HCO_3^- + 53H_2S + 16NH_3 + H_3PO_4$$
$$(CH_2O)_{106}(NH_3)_{16}(H_3PO_4) + 14H_2O \rightarrow 39CO_2 + 14HCO_3^- + 53CH_4 + 16NH_4^+$$
$$+ HPO_4^{2-}$$

(after Canfield and Raiswell 1991)

acids being stronger in the aerobic decomposition, preservation of shells is poor. Shell preservation is maximized in anaerobic conditions, and even the soft tissues may get preserved due to the mineral formation that occurs as a result of the reaction between metal ions and the chemical compounds generated during anaerobic degradation. Bacterial sulphate reduction increases pore-water alkalinity and may enhance calcareous shell preservation. The anaerobic bacteria, occurring in low oxygen environments, are less efficient bio-degraders and may be unable to break down complex biopolymers completely (Allison 2001). The genetic material, DNA, is not so stable as to withstand the vagaries of geological processes. Useful genetic information may not be retained for more than 100 K years, even under the most suitable conditions of rapid burial and cool climate.

In certain environments, the early taphonomic processes may have a profound effect on fossil assemblages. In low marshes, for example, the calcareous tests of foraminifer *Ammonia beccarii* get dissolved, and agglutinated tests with organic cements of *Ammotium salsum* and *Miliammina fusca* are lost because of oxidation of organic matter. As a result, the dead assemblage dominantly comprises robust agglutinated species *Jadammina macrescens* and *Trochammina inflata*, which otherwise form only a small part of the original living assemblage (Martin et al. 2003). The marsh foraminifera are a well-recognized proxy of sea level changes. The differential preservation of life assemblages in this environment may give erroneous signals, but a good understanding of possible taphonomic modifications of foraminiferal assemblages will moderate the problem.

2.3 Post-mortem Transport

While the soft parts are decayed, the shells are also subjected to transportation, fragmentation, dissolution and diagenesis. The microfossils respond differently to these taphonomic processes, depending primarily on mineralogy, size, shape, density and microstructure of the shells. Due to their small size, microfossils are more vulnerable to transportation until they are covered by sediments. Pollen grains are transported by winds over considerable distances. Benthic foraminifera are carried away as suspended particles in macro-tidal estuaries for tens of kilometres up the river mouth until they are deposited on the banks of the river (Ghosh et al. 2009). A quantitative to semi-quantitative assessment of the differential transport of microfossils has been made in flume experiments and field observations. In recent larger benthic foraminifera, the settling velocity and traction velocity (threshold velocity needed to move a grain) vary greatly with size, shape and density of the tests, and the roughness of the sediment surface (Yordanova and Hohenegger 2007). The settling velocity varies from ~4 to 8 cm/s, and the entrainment velocity (on a smooth surface) ranges from ~10 to 16 cm/s (Fig. 2.1). Shape is an important factor in transportation of ostracoda, since the whole carapaces get entrained at a much lower velocity

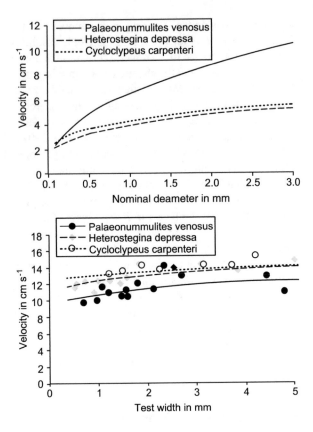

Fig. 2.1 The entrainment velocity (*above*) and settling velocity (*below*) of selected species of modern larger benthic foraminifera (courtesy, Johann Hohenegger)

(6.3–12.7 cm/s) than separate valves (17.5–26.3 cm/s) (Kontrovitz 1975). Flume experiments have shown that traction velocity in foraminifera is related to maximum projection sphericity and the weight of the test (see Box 2.2 for calculation of traction velocity). Martin and Liddell (1991) considered both traction and settling velocities to distinguish groups of foraminifera by their transport potential and found that (1) species having a low settling velocity and low traction velocity have high transport potential (e.g. *Orbulina, Globigerinoides, Sorites, Peneroplis*), and (2) species having a high settling velocity and high traction velocity have low transport potential (e.g. *Discorbis*). Theoretically, an assemblage containing species with similar traction velocities indicates sorting by currents, whereas a large range of values indicates in situ assemblage (Snyder et al. 1990). Bioturbation and other such processes may limit the direct use of the experimental results, but integration with other lines of geological evidence definitely improves the quality of the primary data.

A good example of the use of the basic principles of sediment transport for microfossil accumulation is in the understanding of once extensive carbonate platforms on the margins of the Tethys. The Eocene nummulitic accumulation of these

Box 2.2: Traction Velocity and Shell Transport

The shells of microorganisms while being transported by water are governed by the same basic principles that transport the sediment grains. For a grain to get entrained for transportation, a critical threshold or traction velocity must be reached by the moving fluid. It is a function of fluid velocity, viscosity and particle size, shape and density. Kontrovitz et al. (1978) conducted flume experiments to understand the transport of benthic foraminifera. They found that traction velocity (V_t) is related to maximum projection sphericity (MPS) and weight (W, in mg) of the test by the equation,

$$V_t = 18.4 - 11.4 \text{ MPS} - 38.9 \text{ W} \qquad (1)$$

MPS is the cube root of $S^2/L.I$, where S is the shortest axis of the particle and L and I are the longest and intermediate axes, respectively. They found that mean traction velocities ranged from 5.1 to 18.7 cm/s on a bed of sub-angular to rounded fine sand. Later, the mass term (W) was removed in Eq. (1) and was modified as,

$$V_t = 22.3 - 19.8 \text{ MPS} \qquad (2)$$

Snyder et al. (1990) used the above linear regression to estimate traction velocities of common benthic foraminifera of the Washington continental shelf and distinguished three traction velocity groups—high, intermediate and low—each containing species with similar traction velocities and distinctly different from the other such group. The three groups are as follows:

Group I:	Low traction velocity (<6 cm/s), easily transported
	Nearly equidimensional tests
Group II:	Intermediate traction velocity (6–10 cm/s)
	Intermediate shapes, elongated, inflated or coiled tests
Group III:	High traction velocity (>10 cm/s), most difficult to transport
	Flattened, discoidal and elongated compressed tests

platforms, forming hydrocarbon reservoirs in northern Africa, is a classic example of hydrodynamic sorting of A-form (small-sized) and B-form (large-sized) tests of *Nummulites* (Fig. 2.2). The origin of these accumulations, referred to by many as "nummulitic banks", is debated as to whether they are autochthonous or allochthonous in nature. The experimental measurements of threshold shear velocities indicate that even weak currents can transport large-sized tests of *Nummulites* from their original habitat. Depending on local hydrodynamic conditions, these accumulations may be biocoenosic or transported onshore, foreshore or offshore

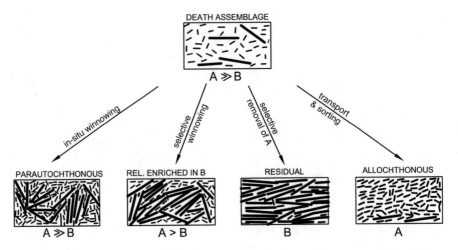

Fig. 2.2 Schematic diagram of hydrodynamic sorting of A- and B- forms of *Nummulites*. Parautochthonous assemblage is dominated by A-forms, allochthonous assemblage consists entirely of A-forms, and the residual assemblages are either enriched in or consist entirely of B-forms (redrawn after Aigner 1985, with permission ©SEPM)

from the original biotope (Fig. 2.3). The paleoenvironmental and paleobathymetric reconstructions of such stratigraphic records need to account for post-mortem changes in the live assemblages.

A number of factors are found to influence death assemblages of planktic foraminifera. These include: (1) ingestion of dead planktic foraminifera by scavengers in the water column and on the seafloor, (2) settling rate, current transport and selective dissolution of tests in the water column and on the seafloor, and (3) scouring and bioturbation (Be 1977). The planktic foraminifera and other microplankton descend to the ocean floor at different rates depending on their shape and density. The settling velocities of modern planktic foraminifera vary from 0.3 to 2.3 cm/s (Berger and Piper 1972).

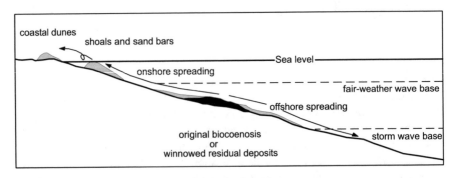

Fig. 2.3 Transportation of *Nummulites* from its original habitat to different environments by waves and currents. The shells may be transported upslope to form shoals, sandbars and coastal dunes, or may spread downslope to offshore (redrawn after Jorry et al. 2006, with permission ©Springer Science+Business Media)

2.4 Diagenesis and Dissolution

Chemical susceptibility to diagenesis and dissolution varies with the composition of the shell. Aragonite is less stable than calcite and, thus, aragonitic shells may be replaced by calcite. Although biogenic structures may remain preserved during calcitization of aragonite, the trace elements and stable isotopes may change markedly, limiting the use of compositional data in paleoenvironmental interpretations. Furthermore, low-Mg calcite (1–4 mol% Mg) is much less soluble than high-Mg calcite (11–19 % Mg). It is due to differential susceptibility that the lysocline (the depth range at which the dissolution rate of calcareous tests increases rapidly) of pteropods is shallower than that of foraminifera. It should be noted that, besides mineralogy, preferential settling of species, size and wall microstructures influence the dissolution behaviour of the shell. According to Be (1977), the more solution-susceptible species (e.g. *Globigerinoides ruber*) are relatively small, thin-walled and have large pores compared with the less solution-susceptible species having large tests, small pores and thick walls (e.g. *Globorotalia tumida*). A general ranking of modern planktic foraminifera is proposed based on their susceptibility to dissolution (Table 2.1). Berger and Piper (1972) attempted to observe whether major dissolution of planktic foraminifera occurs during settling of the test or on the seafloor. It was found that there is no noticeable dissolution in the upper 4000 m of the water column and only the most susceptible species may be affected to some degree during descent. The differential dissolution has important consequences for reconstruction of plankton communities. There may be greater similarity between biocoenosic and thanatocoenosic assemblages of planktic foraminifera in the sediment samples of shallower depths, but only the most resistant species of the life assemblage may be present in the death assemblages at deeper depths (Fig. 2.4).

Table 2.1 Ranking of selected species of planktic foraminifera in order of decreasing susceptibility to solution

Rank	Species	Resistance
1	*Globigerinoides ruber*	Low
2	*Orbulina universa*	
3	*Globigerinoides sacculifer*	
4	*Globigerina bulloides*	
5	*Candeina nitida*	
6	*Globorotalia inflata*	High
7	*Globorotalia menardii*	
8	*Pulleniatina obliquiloculata*	
9	*Sphaeroidinella dehiscens*	
10	*Globorotalia tumida*	

G. ruber (Rank 1) is the least resistant to dissolution and *Gl. tumida* (Rank 10) is the most resistant species in the given assemblage (after Berger 1970)

Fig. 2.4 Selective dissolution of planktic foraminifera during post-mortem settling on the seafloor. Species with low resistance to solution gradually disappear downslope and the assemblage below the lysocline and CCD is represented by the species with high resistance to solution (reproduced after Berger 1985, with permission ©IUGS)

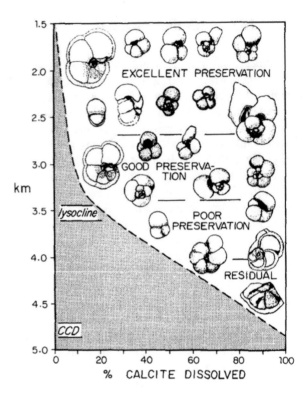

2.5   Time Averaging

Time averaging is the process by which organic remains from different time intervals come to be preserved together (Kidwell 1998). The fossil assemblages are time-averaged due to transport, dissolution, bioturbation and reworking. Even in absence of these factors, time averaging is inherent in fossil assemblages because biological generation times are much shorter than the net rates of sediment accumulation. Microfossils are no exception to this. The duration of time averaging may vary from a few decades to several thousands or millions of years. It is believed that, except in areas with a very high rate of sedimentation (>1 mm/year), the top 1 cm of sediment may represent anything from a decade to hundreds of years of time-averaged foraminiferal accumulation (Murray 2006). Four categories of time-averaged fossil assemblages are recognized, each one having different magnitudes of time averaging depending on the environment (Fig. 2.5). The census assemblage, caused by a mass death of all or part of a community through a catastrophic process,

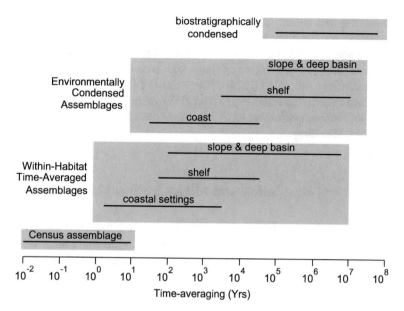

Fig. 2.5 Estimates of time averaging for different types of fossil assemblage (adapted after Kidwell 1998)

has minimal time averaging of individuals. It is like a snapshot of the population. The maximum time-averaged assemblage (tens of millions of years) is a biostratigraphically condensed assemblage, characterized by the species with non-overlapping ages in a single horizon. Most fossil assemblages fall between the two extremes and are categorized as within-habitat time-averaged assemblages and environmentally condensed assemblages.

Intuitively, the taphonomic conditions of shells and the duration of time averaging are likely to be closely related. The well-preserved assemblages, implying a high taphonomic grade, generally reflect accumulation over a short span of time, as compared with poorly preserved assemblages that are, thus, of low taphonomic grade (Brandt 1989). Kidwell (1998) has discussed the possibility of a "taphonomic clock" to determine the relative age of a shell or to estimate the duration of time averaging in a fossil assemblage. Several studies have concluded that there exists, at most, a weak relationship between the scale of time averaging and the taphonomic condition of the shell. The surface condition may just reflect the residence time of the shell at the sediment–water interface. Even though there is no consistent deterioration in taphonomic grades with age, the overall condition of an assemblage degrades with the lapse of time. Modern techniques, including amino acid racemization and AMS ^{14}C dates, have helped in quantitative estimation of time averaging in present-day fossil assemblages. The foraminifera *Buccella* and *Elphidium* in the tidal flat in Mexico are time-averaged at about ~2000 year. A similar order of time averaging is found in the larger foraminifera *Amphistegina gibbosa* and *Archaias angulatus* in carbonate environments. It is estimated that the lower limit of temporal resolution of shallow shelf microfossil assemblage is about 1000 years (Martin et al. 1995, 1996), implying that interpretations at <1000 years resolution are severely constrained in this geologic setting.

2.6   Temporal Resolution and Microstratigraphic Sampling

Sedimentation and taphonomy are the two important aspects for assessment of stratigraphic records in regard to their suitability in unravelling short-term processes. The rate of sedimentation varies widely in different depositional settings and the process of sedimentation may be continuous to intermittent. The sediment preserved in a stratigraphic section represents only a small fraction of the elapsed time. The time represented by hiatuses in a section may be as much as or more than that recorded by actual sediments. It is found that the rate of sedimentation and the period of observation for which the rate is measured are inversely related. This is because, as the period of observation increases, the number and duration of erosions and non-depositions within the observed interval also increases. Schindel (1980) studied the rate and constancy of sedimentation for major sedimentary environments. Fluvial and deltaic environments have high rates of sedimentation, but the sedimentation is intermittent. Deep-sea, inland seas and lakes may have much lower rates of sedimentation, but sediments accumulate continuously. The subjectivity of the plot cannot be ruled out due to extreme variation in both rates and constancy of sedimentation, but as a first order approximation, it is useful for evaluation of possible resolutions of stratigraphic sections. Schindel (1982) proposed "resolution analysis" to estimate the time scale and quality of stratigraphic sequences (see Box 2.3 for the procedure of resolution analysis). The biological and earth processes operate at varying scales of time. Higher temporal resolutions are required for processes with a short time span and, therefore, different scales of sampling, called microstratigraphic sampling, of geological sections are recommended to achieve the required resolution.

The vertical mixing of sediments by organisms (bioturbation) or by waves and currents at the sediment–water interface causes stratigraphic disorder in which the fossils within a stratigraphic sequence are not in proper chronological order. The temporal resolution of a stratigraphic record, as a result, deteriorates. This aspect has been ignored so far due to its complexity. It may not have a marked effect on low-frequency events and low-resolution interpretations, but its consequences are critical for high-resolution interpretations of stratigraphic sections. Various mathematical models of bioturbation have been developed and validated in modern environments.

Box 2.3: Resolution Analysis

Estimation of stratigraphic resolution is becoming increasingly crucial in addressing the issues of evolution, ecology and climate of the deep time. The microstratigraphic sampling (Schindel 1980) for high-resolution biological and environmental processes is challenged by the reality of discontinuous sedimentation. Sediments preserved in the geological record normally represent a fraction of the elapsed time. Schindel (1982) proposed a set of procedures, termed "resolution analysis", to estimate the time scale and quality of sequences. It includes the following:

1. *Temporal scope*, the total span of geologic time encompassed in a sampled sequence. It is estimated by biostratigraphy, radiometric methods or magnetostratigraphy.
2. *Microstratigraphic acuity*, the amount of time represented in each fossiliferous sediment sample. It is estimated from modern short-term sedimentation rate. The sedimentation rate is highly variable between environments and shows an inverse relationship with the time of observation. Short-term rates provide the best acuity, as stratigraphic gaps are minimized.
3. *Stratigraphic completeness*, proportion of the temporal scope represented by actual strata. Completeness can be estimated at any level of resolution.

Plotnick (1986) gave a mathematical model for distribution of hiatuses and estimation of stratigraphic completeness in a given section. He presented the relationship between the "sedimentation rate" vs. the "time span of observation" for different values of stratigraphic gaps, G. The slope of the line is calculated by

$$m = \frac{-\log_{10}(1-G)}{\log_{10}\left[\frac{1-G}{2}\right]} \qquad (1)$$

where G is the relative size of the gap (1/2, 1/3, 1/4, etc.). The stratigraphic completeness is calculated by

$$\text{completeness} = \left[\frac{t}{T}\right]^{-m} \qquad (2)$$

where T is the time span of the whole section and t is the desired short time span.

The profile of radiotracers, including ^{210}Pb (half-life 22.3 years), ^{137}Cs (half-life ~30 years) and ^{234}Th (half-life ~24 days), is used to estimate rate of sedimentation and thickness of the "mixed layer" (in which the mixing of fauna and detrital particles may occur before getting fixed in the historical layer). The thickness of the mixed layer may range up to 1 m in shallow water and 5–10 cm in the deep sea. Berger and Heath (1968) discussed a model for vertical mixing in pelagic sediments to explain how species may get distributed above and below their true levels of appearance and extinction in the sequence (see Box 2.4 for explanation of the mathematical model).

Box 2.4: Mathematical Model of Bioturbation

Berger and Heath (1968) quantitatively described vertical mixing of particles in pelagic sediments. The microplankton and other particles, once deposited on the sediment–water interface, are thoroughly mixed with older deposits in the mixed layer. The particle comes to rest once incorporated into the historical layer (Fig. 2.6). The rate of deposition of sediments determines the rate at which a thin slice at the bottom of the mixed layer is incorporated into the historical layer, thus decreasing the probability of occurrence of the particle in the mixed layer. A small subtraction of sediment layer getting incorporated into the historical layer is expressed as follows:

$$\frac{\text{Probability of being incorporated into historical layer}}{\text{Probability of existence in mixed layer}} = \frac{\text{Thickness of subtracted layer of sediments}}{\text{Thickness of mixed layer}}$$

$$\frac{dP}{P} = -\frac{dL}{m}. \tag{1}$$

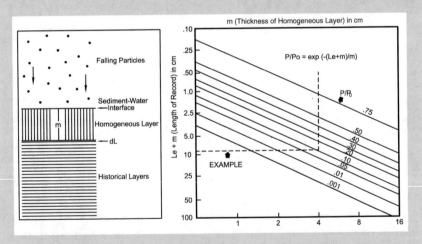

Fig. 2.6 Profile of detrital deep-sea sedimentation explaining homogeneous and historical layers (*left*) and proportions of original concentrations of a species (P/P_0) found in the sediment at a distance of L_e above its level of extinction (redrawn after Berger and Heath 1968, with permission ©author)

(continued)

Box 2.4 (continued)

Integration of Eq. (1) gives the decay formula

$$P = P_o e^{\left(-\frac{L}{m}\right)},$$ (2)

where P is the probability of finding a particle after a thickness L of sediment is deposited on the layer and P_o is the original probability. It implies an exponential decrease in probability of finding a particle with depth in the historical layer.

Equation (2) was used to model the distribution of a species above its level of extinction by

$$P_z = P_{sz} e^{\left[-\frac{(L_e + m)}{m}\right]},$$ (3)

where P_z is the concentration of species z, P_{sz} is the original concentration of z (at a distance of m below the level of extinction), and L_e is the thickness of sediment deposited after the extinction of z. There is a gradual upward decrease in the concentration of a species after its extinction. Based on this model, the proportion of original concentration of a species likely to occur in a core above the extinction datum can be calculated (Fig. 2.6). As illustrated in the figure, 10 % of the original concentration of the species occurs at 9 cm above the true level of extinction. The assumed thickness of the mixed layer is 4 cm. The stratigraphic resolution of deep-sea cores can be calculated with this model. If the rate of sedimentation is 1 cm/k years, the resolution in the given example will be 9000 years. A similar model was proposed for the distribution of species after its appearance:

$$P_z = P_{oz}\left(1 - e^{\left[-\left(\frac{L_a + m}{m}\right)\right]}\right).$$ (4)

The mixing of microfossils changes the radiocarbon age profile of sediments. The un-bioturbated section will show a linear age–depth relationship with zero age at the sediment–water interface. A bioturbated mixed layer will have a constant older age and the particles incorporated into the historical layer carry a record of the mixed layer at the time of final burial. Bioturbation coupled with dissolution may further complicate the radiocarbon age profile in deep-sea records. The age of the mixed layer depends on the balance between sedimentation flux and dissolution flux. The sub-lysocline cores show bias towards a younger age if sedimentation flux > dissolution flux and an older one if sedimentation flux < dissolution flux. In view of this, the ^{14}C stratigraphy and age models of different preservational settings cannot be strictly compared (DuBois and Prell 1988).

References

Aigner T (1985) Biofabrics as dynamic indicators in nummulitic accumulations. J Sediment Petrol 55:131–134

Allison PA (2001) Decay. In: Briggs DEG, Crowther PR (eds) Palaeobiology II. Blackwell, London, pp 270–272

Be AWH (1977) An ecological, zoogeographic and taxonomic review of recent planktonic foraminifera. In: Ramsay ATS (ed) Oceanic micropaleontology, vol 1. Academic, London, pp 1–100

Berger WH (1970) Planktonic foraminifera: selective solution and lysocline. Mar Geol 8:111–138

Berger WH (1985) CO_2 increase and climatic prediction: clues from deep-sea carbonates. Episodes 8(3):163–168

Berger WH, Heath GR (1968) Vertical mixing in pelagic sediments. J Mar Res 26(2):134–143

Berger WH, Piper DJW (1972) Planktonic foraminifera: differential settling, dissolution and redeposition. Limnol Oceanogr 17:275–287

Brandt DS (1989) Taphonomic grades as a classification for fossiliferous assemblages and implications for paleoecology. Palaios 4:303–309

Canfield DE, Raiswell R (1991) Carbonate precipitation and dissolution – its relevance to fossil preservation. In: Allison PE, Briggs DEG (eds) Taphonomy – releasing the data locked in the fossil record. Plenum, New York, pp 411–453

DuBois LG, Prell WL (1988) Effects of carbonate dissolution on the radiocarbon age structure of sediment mixed layers. Deep-Sea Res 35:1875–1885

Ghosh A, Saha S, Saraswati PK, BanerjeeS BS (2009) Intertidal foraminifera in the macrotidal estuaries of the Gulf of Cambay: implications for interpreting sea-level change in palaeo-estuaries. Mar Pet Geol 26:1592–1599

Jorry SJ, Hasler C, Davaud E (2006) Hydrodynamic behaviour of *Nummulites*: implications for depositional model. Facies 52:221–235

Kidwell SM (1998) Time averaging in the marine fossil record: overview of strategies and uncertainties. Geobios 30:977–995

Kontrovitz M (1975) A study of the differential transportation of ostracodes. J Paleontol 49:937–941

Kontrovitz M, Snyder SW, Brown RJ (1978) A flume study of the movement of foraminifera tests. Palaeogeogr Palaeoclimatol Palaeoecol 23:141–150

Martin RE, Liddell WD (1991) Taphonomy of foraminifera in modern carbonate environments: implications for the formation of foraminiferal assemblages. In: Donovan SK (ed) The process of fossilization. Belhaven, London, pp 170–193

Martin RE, Harris MS, Liddell WD (1995) Taphonomy and time-averaging of foraminiferal assemblages in Holocene tidal flat sediments, Bahia la Choya, Sonora, Mexico (northern Gulf of California). Mar Micropaleontol 26:187–206

Martin RE, Wehmiller JF, Harris MS, Liddell WD (1996) Comparative taphonomy of foraminifera and bivalves in Holocene shallow water carbonate and siliciclastic regimes: taphonomic grades and temporal resolution. Paleobiology 22:80–90

Martin RE, Hippensteel SP, Pizzuto JE, Nikitina D (2003) Taphonomy and artificial time-averaging of marsh foraminiferal assemblages (Bombay Hook National Wildlife Refuge, Smyrna, Delaware, USA): implications for rate and magnitudes of late Holocene sea-level change. In: Olson HC, MarkLeckie P (eds) Micropaleontological proxies for sea-level change and stratigraphic discontinuities, vol 75, SEPM Special Publication., pp 31–49

Murray JW (2006) Ecology and applications of benthic foraminifera. Cambridge University Press, Cambridge

Plotnick RE (1986) A fractal model for the distribution of stratigraphic hiatuses. J Geol 94:885–890

Schindel DE (1980) Microstratigraphic sampling and the limits of paleontologic resolution. Paleobiology 6:408–426

Schindel DE (1982) Resolution analysis: a new approach to the gap in the fossil record. Paleobiology 8:340–353

Snyder SW, Hale WR, Kontrovitz M (1990) Assessment of postmortem transport of modern benthic foraminifera of the Washington continental shelf. Micropaleontology 36:259–282

Yordanova EK, Hohenegger J (2007) Studies of settling, traction and entrainment of larger benthic foraminiferal tests: implications for accumulation in shallow marine sediments. Sedimentology 54:1273–1306

Further Reading

Martin RE (1999) Taphonomy: a process approach. Cambridge University Press, Cambridge

Chapter 3
Microfossil Biomineralization and Biogeochemistry

3.1 Introduction

The shells of single-celled micro-organisms contribute significantly to modern-day oceanic sediment. It is estimated that the calcareous-walled foraminifera alone account for ~25 % of the world's carbonate production. Likewise, radiolaria and diatoms are major contributors of siliceous sediment in the deep sea. When more than 30 % of the ocean bottom sediments consist of the skeletal remains of such organisms, it is called an ooze, and these siliceous and calcareous oozes are widespread. The geological record is replete with examples of microfossils as rock builders in the geologic past. Nummulitic limestone, chalk and diatomaceous earth formed primarily of foraminifera, calcareous nannoplankton and diatoms, respectively, are spread over many parts of the world. The major minerals produced by microfossils include calcium carbonate, phosphate and amorphous silica. There are two characteristic features that distinguish the biologically formed minerals from their inorganically produced counterparts. Firstly, the biogenic minerals have unusual external morphologies that develop into intricate and diverse structures. Secondly, these are composite materials consisting of crystals and organic materials (Weiner and Dove 2003).

Organisms construct their shells through precipitation of minerals from the ambient water. The mineralogy of shells and seawater chemistry are, therefore, closely interrelated. Changes in seawater chemistry induce phenotypic changes in the mineralogy of organisms and influence the growth rate of skeletons, and biomineralization apparently alters seawater chemistry that feeds back to influence skeletal mineralogy (Stanley 2006). It is believed that proliferation of siliceous microfossils, radiolaria in the early Paleozoic and diatoms in the late Mesozoic, reduced the concentration of silica in the seawater and prevented siliceous sponges from forming reefs after the Jurassic. It has been shown experimentally that the Mg/Ca ratio of

© Springer International Publishing Switzerland 2016
P.K. Saraswati, M.S. Srinivasan, *Micropaleontology*,
DOI 10.1007/978-3-319-14574-7_3

seawater controls the population, growth rate and skeletal mineralogy of the coc-colithophores. The prolific development of chalk (comprised largely of coccolitho-phores) during the late Cretaceous took place when the seawater Mg/Ca ratio (<1) was at its lowest and Ca concentration (25 to 30 mM) was highest of the Phanerozoic levels (Stanley et al. 2005).

After the pioneering work of Urey et al. (1951) in oxygen and carbon isotope analysis of *Belemnite*, Emiliani (1954, 1955) introduced the method for analysing microfossils so as to know their depth habitats and to interpret the climate of the Pleistocene. Since then, it has become an indispensable tool in paleoceanography and paleoclimatic reconstruction. The trace elements of carbonate shells have like-wise gained importance as tracers of the temperature, salinity, nutrient and chemis-try of seawater. As a corollary, the oxygen and carbon isotope ($\delta^{18}O$, $\delta^{13}C$) and Mg/Ca ratios of microfossil shells provide important information about their paleobiol-ogy, including habitat temperature, reproduction, depth distribution, lifespan, calci-fication and photosymbiosis (Saraswati 2008). Although biogeochemistry has found wide-ranging applications in the ocean and climate sciences, due to notable biologi-cal control in the process of mineralization, a cautious approach should be adopted in the interpretation of shell chemistry.

3.2   Function of the Shell

Why should microfossils have a shell? Brasier (1986) reviewed a number of studies to propose the following functions of mineralized shells in microfossils:

1. The general primary function is protection of the organism.
2. A biomineral shell is an economic way to provide a more rigid protection to the cell wall and organelles than a cellulose or chitin envelope. Silica frustules of diatoms maintain the cell shape and calcareous skeletons of coccolithophorids protect the fragile plasmalemma. The more evolved calcareous wall in foraminifera perhaps expends less energy than the primitive organic and aggluti-nated walls protecting the organelles.
3. It facilitates sinking in non-motile dinoflagellates to remove them rapidly from nutrient depleted surface waters. The motile dinoflagellates are mostly unmineralized.
4. The toxic Ca^{++} has a tendency to enter all cells; the biomineralization pumps it out. Expulsion of Ca^{++} is necessary just prior to gametogenesis in planktic fora-minifera. A large sac-like final chamber in *Globigerinoides sacculifer* is believed to perform this function.

3.3 Processes of Biomineralization

There are two principal processes of biomineralization: biologically induced and biologically controlled. Environment plays a major role in the biologically induced process. Photosynthetic organisms, for example, induce calcium carbonate precipitation by consuming CO_2. In biologically controlled mineralization, cellular activities control the mineralogy of the skeletons. A localized zone having and maintaining a sufficient super-saturation is required for mineral precipitation. In most biological systems, the site of mineral deposition is isolated from the environment so that this region delimits diffusion into and out of the system. Intracellular vesicles create such a compartmentalized environment where compositions can be regulated (Weiner and Dove 2003).

A vast majority of microfossils have calcareous and siliceous walls. The process of biomineralization is better studied in foraminifera owing to the importance of the group in stratigraphy and environment. The mineralized structures are of fundamental value in the classification and identification of foraminifera. The foraminifera progressively evolved from their naked ancestors into organically walled, agglutinated, calcareous imperforate (miliolid) and calcareous perforate (rotaliid) shell-wearing organisms. Calcification in foraminifera is intracellular, and the two major groups, miliolid and rotaliid, calcify differently. In the rotaliid (or hyaline) type, the inorganic carbon and calcium are stored in separate intracellular pools for extracellular precipitation of calcium carbonate (ter Kuile et al. 1989). A primary organic sheet in the shape of the new chamber is produced before calcification. It provides the site of nucleation for calcite crystals where the crystals are arranged with c-axes perpendicular to the wall. The miliolid (or porcelaneous) foraminifera do not have intracellular pools, but they obtain inorganic carbon directly from seawater. Calcite is precipitated in the form of needles within the cytoplasmic vesicles. At the time of chamber formation, the needles are transported outside the shell and get randomly deposited onto the organic matrix, unlike the preferentially oriented crystals in the rotaliid type (Fig. 3.1). It is further observed in the rotaliid foraminifer *Amphistegina lobifera* that an Mg-rich primary calcite is first precipitated as microspherules with an organic matrix at the boundary between ectoplasm and endoplasm. This is followed by the deposition of low-Mg secondary calcite, forming 90 % of the total mass (Erez and Bentov 2002). The modern techniques of atomic force microscopy and synchrotron-based fluorescence mapping have given insights into the sub-micrometre scale structures of the shell. The organic matter (protein and polysaccharides) is seen to be permanently incorporated within the calcified structure at the nanometre scale, contradicting the template model for calcification in foraminifera (Cuif et al. 2011) that envisages deposition of calcite on the organic lining. The surface seawater is supersaturated with calcite, but a high concentration of Mg does

Fig. 3.1 Calcification in hyaline and porcelaneous foraminifera: hyaline foraminifera store inorganic carbon and calcium in separate intracellular pools for calcification (**a**), but porcelaneous foraminifera (**b**) obtain carbon for calcification directly from the seawater and precipitate calcite in vesicles (redrawn after ter Kuile et al. 1989, with permission ©Springer Science + Business Media)

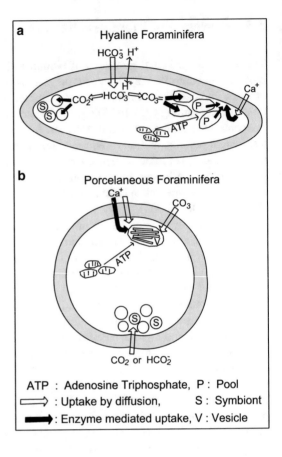

not allow its precipitation. Foraminifera actively remove Mg from vacuolized seawater before precipitation. It is shown experimentally that foraminifera are able to elevate pH at the site of calcification by at least one unit above the ambient seawater pH, and thereby overcome precipitation inhibition (de Nooijer et al. 2009). Some benthic and planktic foraminifera harbour algal symbionts. The culture experiments on these species have shown high growth rate and higher calcium incorporation with an increase in light intensity, suggesting that the photosynthetic activity of the symbionts stimulates calcification (Box 3.1 explains how the rate of calcification varies between day and night in symbiont-bearing larger benthic foraminifera). The chemical microenvironment adjacent to the shell surface of these foraminifera is also quite different from the seawater and, as a result, it has a distinct effect on isotopic composition and on certain trace element concentrations in the shell (see

Box 3.1: Role of Photosymbiosis in Calcification

Symbionts play an important role in the calcification of symbiont-bearing benthic and planktic foraminifera. Rottger et al. (1980) noted that the growth rates of *Heterostegina depressa* and *Amphistegina lessonii* increased with the intensity of light, and that the optimal light intensity for growth was 400–800 lx. Growth stopped in the dark, but it continued, even in the absence of food, when incubated in light. In another experiment, Duguay (1983) measured calcium incorporation and carbon fixation by microalgal symbionts in *Archaias angulatus*, *Sorites marginalis* and *Cyclorbiculina compressa*. It was found that the rate of calcification was 2–3 times higher in the light (when symbiont photosynthesis was active) than in the dark (Fig. 3.2).

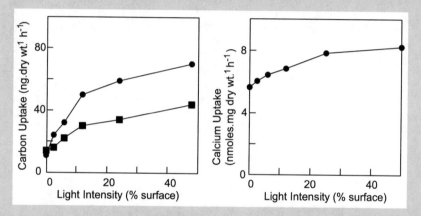

Fig. 3.2 The effect of light intensity on carbon and calcium intake by a larger benthic foraminifer *Archaias angulatus*. The organic and inorganic carbon uptake (*left*) are indicated by the *circle* and the *rectangle*, respectively (redrawn after Duguay 1983, with permission ©Cushman Foundation for Foraminiferal Research)

Box 3.2 for results of micro-sensor studies on changes in the chemical profile in proximity to the shell).

Another common group of calcareous microfossils is the coccoliths, which are usually composed of low-Mg calcite. Two main types of coccolith are recognized. The holococcoliths are made of numerous simple calcite crystals of uniform size held together by an organic matrix, and heterococcoliths have a diverse morphology of calcite crystals, including rim, plate and tubes. The coccoliths are formed in vesicles issuing from Golgi apparatus, where a thin organic base plate is formed for calcite crystal nucleation. Calcification starts around the rim of the plate and grows to form single crystalline elements that constitute the complete coccolith (Faber and Presig 1994).

Box 3.2: Chemical Microenvironments Around Symbiont-Bearing Foraminifera

In symbiont-bearing foraminifera, a distinct microenvironment is created in the vicinity of the test due to calcification, respiration and symbiont photosynthesis. Micro-sensor measurements have indicated that it is characterized by a dynamic change in the value of pH $(=-\log_{10}[H^+])$ and concentrations of CO_2 and O_2 between day and night (Fig. 3.3). Carbonate chemistry (CO_3^{2-} and HCO_3^-) in the vicinity of the test, as a result, is quite different from that of the seawater (Köhler-Rink and Kühl 2000, 2005). The CO_2 fixation by photosynthesis is expressed as follows:

$$6H_2O + 6CO_2 + \text{sunlight} \rightarrow 6O_2 + C_6H_{12}O_6.$$

The source of CO_2 in the above reaction is the dissolved CO_2 present in the seawater existing in the following state of equilibrium:

$$H_2O + CO_2 \leftrightarrow H_2CO_3$$

As CO_2 is consumed during photosynthesis, equilibrium shifts from right to left, decreasing the concentration of carbonic acid. As a result, there is an

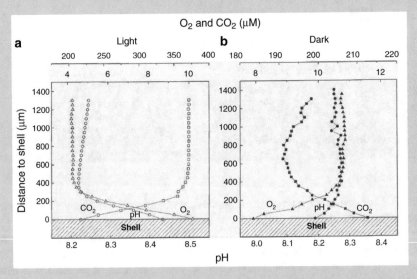

Fig. 3.3 Concentration profiles of O_2, CO_2 and pH under light (A) and dark (B) conditions in the vicinity of the larger benthic foraminifer *Marginopora vertebralis*. Profiles are measured from shell surface towards the well-mixed surrounding seawater (reproduced after Köhler-Rink and Kühl 2000, with permission ©Springer Science + Business Media)

(continued)

Box 3.2 (continued)

increase in pH (decrease in hydrogen ion concentration). Due to the increase in pH, the concentration of HCO_3^- and CO_3^{2-} increases in the following equilibria:

$$H_2CO_3 \leftrightarrow H^+ + HCO_3^-,$$
$$HCO_3^- \leftrightarrow H^+ + CO_3^{2-}$$

The rise in pH and subsequent rise in CO_3^{2-} increases the degree of $CaCO_3$ saturation, which can lead to enhanced calcium carbonate precipitation. In the dark, the level of CO_2 respired by foraminifera builds up in the absence of photosynthesis. The value of the pH is lowered and the concentration of HCO_3^- and CO_3^{2-} decreases. The metabolic process of foraminifera affects the $\delta^{13}C$ of the precipitating calcium carbonate due to preferential use of $^{12}CO_2$ in photosynthesis by the symbionts and the ^{13}C depleted respiratory CO_2 of the host.

Diatoms are unicellular microalgae having cells covered by a siliceous "frustule". Silicon for frustule formation is taken from the environment in a soluble form as $Si(OH)_4$ (mono silicic acid). The silica formation process occurs in a silica deposition vesicle (SDV), which is a modified Golgi apparatus (Fig. 3.4). Organic molecules, peptides (silaffins) and polyamines in the SDV control the deposition of silica. The frustule of a diatom is made of two petri-dish-like valves (the epitheca, the

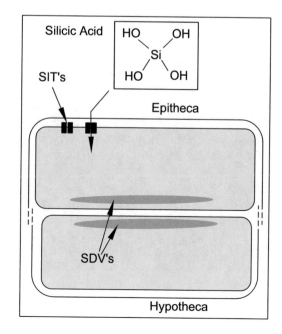

Fig. 3.4 Silica biomineralization in a diatom. The frustules are formed in the silica deposition vesicle (SDV). The frustule is made of a larger valve, epitheca, and a smaller valve, hypotheca (redrawn after Brunner et al. 2009, with permission ©Springer Science + Business Media)

larger one at the top, and the hypotheca, the smaller one at the bottom). A number of girdles are placed between the two. During cell division, each sibling cell produces a valve SDV which grows with the deposition of silica. After completion of the two valves, they are exocytosed and the two daughter cells become fully independent. An organic membrane on the external surface of the frustule protects it from direct contact with water to prevent dissolution of hydrated silica (Cuif et al. 2011).

3.4 Oxygen and Carbon Isotopes in Foraminifera

Geologists have long inferred paleoclimate on the basis of ancient sediments and the associated fossils and given a qualitative estimate of a change in the climate. The development of the isotopic technique made it possible to quantify the change. The basis of the isotopic paleothermometer is temperature dependence of the fractionation of the two isotopes of oxygen (O^{16} and O^{18}) between calcite and water (see Box 3.3 for

---

Box 3.3: Isotope Fractionation

When calcite is precipitated from seawater, either inorganically or by organisms constructing their shells, the ratio of "heavy" to "light" isotopes gets distributed differently from one phase (water) to the other (calcite). Such a partitioning of isotopes between two phases A and B is referred to as the isotopic fractionation factor and is defined as,

$$\alpha_{A-B} = R_A / R_B,$$

where R refers to the atomic ratio of a heavy to a light isotope ($^{18}O/^{16}O$ or $^{13}C/^{12}C$). The thermodynamically calculated α (equilibrium fractionation factor) for calcite–water is 1.0288 at 25 °C. When the measured value of α is equal to the equilibrium fractionation factor for the particular temperature, the precipitation is said to be in isotopic equilibrium; otherwise, it is disequilibrium precipitation. Biogenically precipitated calcite generally has disequilibrium isotopic fractionation. The degree of disequilibrium is generally smaller for oxygen than for carbon, as there is a larger pool of oxygen (water) and a smaller one of carbon (dissolved bicarbonate) interacting with the metabolic products during shell formation (Wefer and Berger 1991). In isotopically derived paleotemperature estimation, it is necessary to ensure that the equilibrium precipitation of the oxygen isotope or the magnitude of disequilibrium be known so as to make the appropriate correction.

an explanation of isotope fractionation and Box 3.4 for isotope notation). Historically, isotope fractionation and the use of oxygen isotopic ratios as a geothermometer were pioneered by Urey (1947, 1951). McCrea (1950) experimentally determined the relationship between temperature and oxygen isotopic composition of

---

Box 3.4: Isotope Notations

Isotopes are atoms whose nuclei have the same number of protons but a different number of neutrons. Stable isotopes are non-radioactive, and there are about 300 such isotopes in nature. In micropaleontology, the application of stable isotopes is almost exclusively focused on oxygen and carbon (Table 3.1). In isotopic analyses, the absolute abundance of a minor isotope (e.g. ^{18}O or ^{13}C) or the absolute values of isotopic ratio (e.g. $^{18}O/^{16}O$) are difficult to measure with sufficient accuracy. But the difference in absolute isotopic ratios between two substances (say, sample and "standard") can be measured directly through isotope ratio mass spectrometry with far more precision. The difference in the oxygen isotopic ratios of the sample and the standard ($\delta^{18}O$) is defined as follows and expressed in per mill (‰) unit:

$$\delta^{18}O = \left[\left\{ \left(^{18}O/^{16}O \right)_{sample} - \left(^{18}O/^{16}O \right)_{standard} \right\} / \left(^{18}O/^{16}O \right)_{standard} \right] \times 1000.$$

The carbon isotopic composition is similarly expressed as,

$$\delta^{13}C = \left[\left\{ \left(^{13}C/^{12}C \right)_{sample} - \left(^{13}C/^{12}C \right)_{standard} \right\} / \left(^{13}C/^{12}C \right)_{standard} \right] \times 1000$$

A positive δ value indicates enrichment in the heavy isotopes, relative to the standard, and a negative value indicates depletion.

For oxygen and carbon isotopes in carbonates, *Belemnitella americana* from the Peedee Formation in North Carolina, USA (abbreviated as PDB), is used as "standard". Since the original PDB standard is no longer available, various other international standards calibrated with PDB are available and distributed by the National Institute of Standards and Technology (NIST, USA) and the International Atomic Energy Agency (IAEA, Vienna). SMOW (standard mean ocean water) is used as the standard for water samples. The principles of isotope measurements and standards are discussed by Hoefs (2009).

Table 3.1 The stable isotopes of oxygen and carbon and their relative abundance

Isotopes		Abundance (%)
Oxygen	^{16}O	99.7630
	^{17}O	00.0375
	^{18}O	00.1905
Carbon	^{12}C	98.98
	^{13}C	01.11

inorganically precipitated calcium carbonate, and later, Epstein et al. (1951) investigated the same relationship in marine mollusks. Craig (1965) modified the equation of Epstein et al. (op cit) as follows to determine temperature from the oxygen isotopic value of shell carbonate:

$$T\left(°C\right) = 16.9 - 4.2\left(\delta_c - \delta_w\right) + 0.13\left(\delta_c - \delta_w\right)^2,$$

where δ_c is $\delta^{18}O$ of CO_2 obtained from the carbonate by reaction with phosphoric acid at 25 °C with respect to a mass spectrometer working standard gas and δ_w is $\delta^{18}O$ of CO_2, equilibrated isotopically at 25 °C with the water from which the carbonate was precipitated, measured against the same working standard.

There are three essential requirements for isotopic paleothermometry: (1) preservation of the original isotopic composition in the shell, (2) precipitation of $CaCO_3$ in close isotopic equilibrium with the surrounding seawater, and (3) estimation of δ_w of the water in which the shell calcified. Samples are examined under a microscope to ensure their pristine preservation. They are first examined under an optical microscope for their surface features and clear primary wall mineralogy free of iron stain and chamber infillings by secondary minerals. Cathodoluminescence microscopic observation can reveal the presence of diagenetic minerals in the sample. The optically pristine samples are then examined under a scanning electron microscope to see that the shell microstructures are clearly preserved and there is no overgrowth of secondary minerals. Another essential requirement is that the shell calcite should have been precipitated in close isotopic equilibrium with the surrounding seawater. Not all organisms secrete their shell in isotopic equilibrium with the seawater. The mollusks secrete their shells in isotopic equilibrium, but not the echinoids, crinoids or corals (McConnaughey 1989, Wefer and Berger 1991). Most paleoclimate studies are based on isotopic analysis of foraminifera. Some species of foraminifera grow in isotopic equilibrium with seawater, while others grow out of equilibrium (or under a disequilibrium condition). The disequilibrium effect is due to the influence of factors such as respiration, symbiont photosynthesis, gametogenic calcite and carbonate ion concentration, the combined effects of which are generally known as the "vital effect" (see Rohling and Cooke 1999 for a detailed review). If the magnitude of disequilibrium is constant and known, the isotopic values of the sample can suitably be corrected for paleotemperature calculation. The third requirement of estimation of δ_w is not an easy parameter to establish with certainty. The $\delta^{18}O$ of water is related to salinity and the amount of water locked up in continents in the form of glacial ice (strongly enriched in ^{16}O). Oceans, being generally well-mixed large reservoirs, are buffered against significant fluctuations in their isotopic compositions. At present, the maximum observed difference between various parts of the major oceans is 1.4‰. Shackleton (1984) estimated that the $\delta^{18}O$ of oceans changed by 1.0–1.4‰ during the glacial–interglacial periods of the Pleistocene.

Table 3.2 $\delta^{18}O$ and $\delta^{13}C$ disequilibrium correction factor for selected species of foraminifera

S.No.	Species	$\delta^{18}O$ ‰ (PDB)	$\delta^{13}C$ ‰ (PDB)	Reference
1	*Marginopora kudakajimaensis*	−0.3	−2.1	Saraswati et al. (2004)
2	*Amphisorus hemprichii*	−0.1	−1.4	Saraswati et al. (2004)
3	*Amphistegina lessonii*	+0.3	−4.0	Saraswati et al. (2004)
4	*Oridorsalis spp.*	0.00	−0.90	Zachos et al. (1994)
5	*Nuttallides*	−0.40	0.00	Zachos et al. (1994)
6	*Cibicidoides spp.*	−0.60	0.00	Zachos et al. (1994)
7	*Uvigerina spp.*	0.00	−0.90	Zachos et al. (1994)
8	*Globigerina bulloides*	0 to +0.5	–	Niebler et al. (1999)
9	*Globorotalia menardii*	−0.2	–	Niebler et al. (1999)
10	*Globigerinoides ruber*	0 to −1.0	–	Niebler et al. (1999)

The isotope paleothermometer has seen three important refinements: (1) determining disequilibrium correction factor, (2) paleolatitude-related $\delta^{18}O$ correction of seawater and (3) species-specific paleotemperature equations. It is of critical importance to ascertain whether the species analysed secreted its shell in equilibrium or deviated from equilibrium by a known amount. The disequilibrium correction factors have been suggested for some of the commonly used foraminiferal species (Table 3.2). In the modern ocean, the surface water $\delta^{18}O$ varies with latitude, which is expressed as follows:

$$Y = 0.576 + 0.041X - 0.0017X^2 + 1.35 \times 10^{-5} X^3$$

where Y is the $\delta^{18}O$ of water and X is the absolute latitude (Zachos et al. 1994). In paleotemperature determination, therefore, instead of a global value of δ_w, a site-specific δ_w can be calculated for the known paleolatitudinal position of the investigated area.

Due to varying magnitude of the vital effect, species-specific paleotemperature equations are proposed and used for better results. Erez and Luz (1983) proposed the following equation based on laboratory-cultured *Globigerinoides sacculifer*:

$$T(^{\circ}C) = 17.0 - 4.52(\delta_c - \delta_w) + 0.03(\delta_c - \delta_w)^2.$$

An extensive review of various aspects of oxygen isotope calibration is given in Bemis et al. (1998).

Table 3.3 Depth ranking of planktonic foraminifera based on plankton tow and oxygen isotope data

Plankton tow data	Oxygen isotope data
0–150 m	0–50 m
G. ruber, G. sacculifer, G. conglobatus	*G. ruber, G. sacculifer, G. conglobatus*
50–100 m	0–100 m
O. universa, N. dutertrei,	*N. dutertrei*
P. obliquiloculata, G. menardii	50– >100 m
>150 m	*P. obliquiloculata, G. menardii,*
G. bulloides, C. nitida, S. dehiscens,	*G. truncatulinoides*
G. tumida, G. truncatulinoides	

Compiled from Hecht 1976

The stable isotopes of oxygen and carbon have important applications in the paleoecology and paleobiology of microfossils. Emiliani (1954) used the oxygen isotopic composition of planktic foraminifera to infer that *Globigerinoides ruber* and *Globigerinoides sacculifer* lived close to surface waters and *Globorotalia* lived in deeper waters. Several studies have shown that the isotopic depth rankings of planktic foraminifera are in good agreement with the depth stratification data of plankton tows (Table 3.3). An analysis of Miocene planktic foraminifera from the Pacific has shown that relative depth rankings for most species did not change except for two, *Globorotalia menardii* and *Globorotalia limbata*, that changed from deep to intermediate and from deep to shallow, respectively (Gasperi and Kennett 1992). The $\delta^{18}O$ and $\delta^{13}C$ studies on fossil planktic foraminifera across the geologic time, however, suggest that the modern analogue of keeled globorotalids inhabiting deeper water compared with unkeeled globigerines is oversimplified. *Globoquadrina* in the Neogene and *Catapsydrax* in the Oligocene were deep-water taxa. In the Paleogene, the globorotalid-like genera *Morozovella* and *Acarinina* were surface dwellers, while the globigerine-like *Subbotina* lived in deep water (Corfield and Cartlidge 1991).

Several living benthic and planktic foraminifera host symbiotic algae, as it helps them obtain energy in oligotrophic environments and promote calcification. A similar symbiotic relationship is believed to have existed in some foraminifera of the geologic past. But there is no direct evidence of photosymbiosis in the fossil species. Due to a marked difference in isotopic composition between symbiont-bearing and symbiont-free foraminifera, the isotopic method is recognized as a potential tool for inferring photosymbiosis in fossil taxa (see Box 3.5 for isotopic characteristics of photosymbiosis).

Box 3.5: Isotopic Evidence of Photosymbiosis in Foraminifera

Photosynthesis is unlikely to have any direct effect on ^{18}O depletion, although in some culture experiments, the symbiont-bearing planktic foraminifera show a decrease in $\delta^{18}O$ with the intensity of light. It does, however, have a distinct influence on the carbon isotopic composition of the shell. Photosynthesis preferentially removes ^{12}C and respiration produces CO_2, which is enriched in ^{12}C. The carbon isotopic composition of the shell will depend on the relative contributions of respired and photosynthetically produced carbon flux during calcification. The oxygen and carbon isotopic composition of the symbiont-bearing planktic and benthic foraminifera has been studied to evolve criteria for inferring photosymbiosis in fossil foraminifera (Table 3.4). In an assemblage of Maastrichtian planktic foraminifera, *Planoglobulina acervulinoides* and *Racemiguembelina fruticosa* were inferred to be photosymbiotic due to strongly negative $\delta^{18}O$ and poor correlation between $\delta^{18}O$ and $\delta^{13}C$. *Planoglobulina multicamerata* in the same assemblage, having most positive $\delta^{18}O$ values and a strong co-variance between size-related $\delta^{18}O$ and $\delta^{13}C$ values, was interpreted to have lacked photosymbionts.

Table 3.4 Oxygen and carbon isotopic evidence of photosymbiosis in planktic and benthic foraminifera

Isotopic characteristics	Planktic (Houston et al. 1999)	Benthic (Saraswati et al. 2004)
Value of $\delta^{18}O$ as compared with co-existing asymbiotic taxa	More negative	In the same range
Size-related change in $\delta^{18}O$	Small	Large for shallow-water species to small for deep-water species
Size-related change in $\delta^{13}C$	Large	Large
$\delta^{18}O$–$\delta^{13}C$ correlation	Poor	Poor
Intra-specific variability in $\delta^{13}C$	Significant	Significant

3.5   Trace Elements in Foraminifera

The minor and trace elements incorporated into foraminiferal shells at the time of calcification are proven paleo-proxies of climate, nutrient and seawater composition. Ca^{2+} ion in the calcitic shell is replaced by the divalent cations, including Mg^{2+}, Sr^{2+}, Mn^{2+}, Cd^{2+} and Ba^{2+}. The empirical partition coefficients (D) and the application of these trace elements are shown in Table 3.5. The incorporation of Mg into calcite is dependent on the temperature of the surrounding seawater during growth such that foraminiferal Mg/Ca ratios increase with increasing temperature. The Mg/Ca thermometry is, thus, a new addition to the expanding list of paleotemperature proxies. Most studies

Table 3.5 Trace elements in foraminiferal calcite (their abundance in foraminifera and seawater, partition coefficient and proxy)

| Element | Concentration in foraminifera | | Seawater concentration (mol/Kg) | Partition coefficient (D) | Proxy |
	Planktonic (mol/mol Ca)	Benthic (mol/ mol Ca)			
Magnesium	$0.5–5 \times 10^{-3}$	$0.5–10 \times 10^{-3}$ (neritic = 0.13– 0.15)	53.2×10^{-3}	$0.1–1 \times 10^{-3}$	Temperature
Strontium	$1.2–1.6 \times 10^{-3}$	$0.9–1.6 \times 10^{-3}$ (neritic = 0.13– 0.15)	90×10^{-6}	0.11–0.19	Seawater chemistry
Barium	$0.5– 2(10) \times 10^{-6}$	$1.5–5 \times 10^{-6}$	$32–150 \times 10^{-9}$	0.15–0.4	Alkalinity
Cadmium	$0.002– 0.1 \times 10^{-6}$	$0.02– 0.25 \times 10^{-6}$	$0.001– 1.1 \times 10^{-9}$	1.3–3.0	Phosphate (nutrient)

The partition coefficient is the ratio of metal/Ca in the foraminifera shell to metal/Ca in the seawater (after Lea 1999)

have shown foraminiferal Mg/Ca to be related exponentially to seawater temperature and a number of species-specific calibrations have been proposed (Table 3.6).

The exact nature of the influence of other factors, including seawater properties and physiology of the organisms, is still a subject of ongoing research. The cultured specimens of planktic foraminifera *Globigerinoides sacculifer* and *Orbulina universa* and several larger foraminifera show no correlation between Mg/Ca ratio and seawater temperature (Delaney et al. 1985; Raja et al. 2005, 2007; Segev and Erez 2006). High-resolution analysis by laser ablated sampling of single chambers of *Orbulina universa* has revealed the presence of multiple low- and high-Mg bands formed due to diurnal variation in pH within the foraminiferal microenvironment and not due to the temperature of the seawater (Eggins et al. 2004). Furthermore, due to changes in Mg/Ca of the past oceans by tectonic processes and biological evolution, the use of a Mg/Ca thermometer has some degree of uncertainty in regard to longer timescales. Notwithstanding these limitations, it is a useful

Table 3.6 The Mg/Ca–Temperature calibrations for selected species of benthic foraminifera

| Species | T Range (°C) | Mg/Ca = B × exp (A × T) | | Sample type | References |
		B	A		
Cassidulina neoteretis	0.96– 5.47	0.864 (±0.07)	0.082 (±0.020)	Surface samples	Kristjansdottir et al. (2007)
Cibicidoides spp.	0.8– 18.4	0.867 (±0.049)	0.109 (±0.007)	Core tops	Lear et al. (2002)
Marginopora kudakajimaensis	21–29	143.18	0.0317	Alive	Raja et al. (2005)
Multiple species of planktic foraminifera	–	0.38	0.090	Sediment trap	Anand et al. (2003)
Orbulina universa	16–27	1.38	0.085	Culture	Lea et al. (1999)

proxy for paleotemperature when combined with the $\delta^{18}O$ of the same phase to correct the effects of $\delta^{18}O_{water}$ variation during glacial–interglacial intervals.

Wide-ranging applications have also been found for the trace elements in other aspects of oceanography. Cadmium and barium are used as proxies for nutrient. The ratio of boron isotopes (^{10}B and ^{11}B) denoted by $\delta^{11}B$ is a proxy for pH and shows a major shift between glacial and interglacial periods. The isotopes of Sr and Nd record changes in ocean chemistry. Secular variation in the $^{87}Sr/^{86}Sr$ ratio is a well-established stratigraphic tool. Understanding the relationship between shell chemistry and seawater composition and physical and biological factors is extremely challenging, but with a precise knowledge of the controlling factors and in conjunction with stable isotopes, foraminiferal trace elements should provide a powerful tool for studying ocean and climate evolution (see Lea 1999 for a detailed review).

References

Anand P, Elderfield H, Conte MH (2003) Calibration of Mg/Ca thermometry in planktonic foraminifera from a sediment trap time series. Paleoceanography 18, Doi: 10.1029/2002PA000846

Bemis BE, Spero HJ, Bijma J, Lea DW (1998) Reevaluation of the oxygen isotopic composition of planktonic foraminifera: Experimental results and revised paleotemperature equation. Paleoceanography 13(2):150–160

Brasier M (1986) Why do lower plants and animals biomineralize? Paleobiology 12(3):241–250

Brunner E, Groger C, Lutz K, Richthammer P, Spinde K, Sumper M (2009) Analytical studies of silica biomineralization: towards an understanding of silica processing by diatoms. Appl Microbiol Biotechnol 84:607–616

Corfield RM, Cartlidge JE (1991) Isotopic evidence for the depth stratification of fossil and recent Globigerinina: a review. Hist Biol 5:37–63

Craig H (1965) The measurement of oxygen isotope paleotemperatures. In: Tongiorgi E (ed) Stable isotopes in oceanographic studies and paleotemperatures: Spoleto 1965. Consiglio Nazionaledella Ricerche Laboratorio di Geologia Nucleare, Pisa, pp 161–182

Cuif J, Dauphin Y, Sorauf JE (2011) Biominerals and fossils through time. Cambridge University Press, Cambridge

Delaney ML, Be AWH, Boyle EA (1985) Li, Sr, Mg, and Na in foraminiferal calcite shells from laboratory culture, sediment traps, and sediment cores. Geochim Cosmochim Acta 49:1327–1341

Duguay LE (1983) Comparative laboratory and field studies on calcification and carbon fixation in foraminiferal-algal associations. J Foraminifer Res 13:252–261

Eggins SM, Sadekov A, De Deckker P (2004) Modulation and daily banding of Mg/Ca in *Orbulinauniversa* tests by symbiont photosynthesis and respiration: a complication for seawater thermometry? Earth Planet Sci Lett 225:411–419

Emiliani C (1954) Depth habitats of some species of pelagic foraminifera as indicated by oxygen isotope ratios. Am J Sci 252:149–158

Emiliani C (1955) Pleistocene temperatures. J Geol 63:538–578

Epstein S, Buchsbaum R, Lowenstam HA, Urey HC (1951) Carbonate-water isotopic temperature scale. Geol Soc Am Bull 62:417–426

Erez J, Bentov S (2002) Calcification processes in foraminifera and their paleo-oceanographic implications. Abs. Forams 2002, International Symposium on Foraminifera, Perth, pp 32–33

Erez J, Luz B (1983) Experimental paleotemperature equation for planktonic foraminifera. Geochim Cosmochim Acta 47:1025–1031

Faber WW, Presig HR (1994) Calcified structures and calcification in protists. Protoplasma 181:78–105

Gasperi JT, Kennett JP (1992) Isotopic evidence for depth stratification and paleoecology of Miocene planktonic foraminifera: Western Equatorial Pacific DSDP site 289. In: Tsuchi R, Ingle JC Jr (eds) Pacific Neogene – environment, evolution and events. University of Tokyo Press, Tokyo, pp 117–147

Hecht AD (1976) The oxygen isotopic record of foraminifera in deep sea sediment. In: Hedley RH, Adams CG (eds) Foraminifera, 2nd edn. Academic, London, pp 1–43

Hoefs J (2009) Stable isotope geochemistry. Springer, Heidelberg

Houston RM, Huber BT, Spero HJ (1999) Size-related isotopic trends in some Maastrichtian planktic foraminifera: methodological comparisons, intra-specific variability and evidence for photosymbiosis. Mar Micropaleontol 36:169–188

Köhler-Rink S, Kühl M (2000) Microsensor studies of photosynthesis and respiration in larger symbiotic foraminifera: I The physico-chemical microenvironment of *Marginopora vertebralis, Amphistegina lobifera* and *Amphisorus hemprichii*. Mar Biol 137:473–486

Köhler-Rink S, Kühl M (2005) The chemical microenvironment of the symbiotic planktonic foraminifer *Orbulina universa*. Mar Biol Res 1:68–78

Kristjansdottir GB, Stoner JS, Jennings A, Andrews JT, Gronvold K (2007) Geochemistry of Holocene cryptotephras from the North Iceland Shelf (MD99- 2269), Inter calibration with radiocarbon and paleomagnetic chronostratigraphies. The Holocene 17(2):155–157

ter Kuile B, Erez J, Padan E (1989) Mechanisms for the uptake of inorganic carbon by two species of symbiont-bearing foraminifera. Mar Biol 103:73–90

Lea DW (1999) Trace elements in foraminiferal calcite. In: Sen Gupta BK (ed) Modern Foraminifera. Kluwer Academic, Dordrecht, The Netherlands, pp 259–277

Lea DW, Mashiotta TA, Spero HJ (1999) Controls on magnesium and strontium uptake in planktonic foraminifera determined by live culture. Geochim Cosmochim Acta 63:2369–2379

Lear CH, Rosenthal Y, Slowey N (2002) Benthic foraminiferal paleothermometry: a revised coretop calibration. Geochim Cosmochim Acta 66:3375–3387

McConnaughey T (1989) ^{13}C and ^{18}O isotopic disequilibrium in biological carbonates: I. Patterns. Geochim Cosmochim Acta 53:151–162

McCrea JM (1950) On the isotopic chemistry of carbonates and a paleotemperature scale. J Chem Phys 18:849–857

Niebler HS, Hubberten HW, Gersonde R (1999) Oxygen isotope values of planktic foraminifera: a tool for reconstruction of surface water reconstruction. In: Fischer G, Wefer G (eds) Use of proxies in paleoceanography: examples from the South Atlantic. Springer, Berlin, pp 165–189

deNooijer LJ, Toyofuku T, Kitazato H (2009) Foraminifera promote calcification by elevating their intracellular pH. Proc Natl Acad Sci U S A 106(36):15374–15378

Raja R, Saraswati PK, Rogers K, Iwao K (2005) Magnesium and strontium compositions of recent symbiont-bearing benthic foraminifera. Mar Micropaleontol 58:31–44

Raja R, Saraswati PK, Iwao K (2007) A field-based study on variation in Mg/Ca and Sr/Ca in larger benthic foraminifera. Geochem Geophys Geosyst 8, Q10012. doi:10.1029/200 6GC001478

Rohling EJ, Cooke S (1999) Stable oxygen and carbon isotopes in foraminiferal carbonate shells. In: Sen Gupta BK (ed) Modern Foraminifera. Kluwer, Dordrecht, The Netherlands, pp 239–258

Rottger R, Irwan A, Schmaljohann R, Franzisket L (1980) Growth of the symbiont bearing foraminifera *Amphistegina lessonii* d'Orbigny and *Heterostegina depressa* d'Orbigny (Protozoa). In: Schwemmler W, Schenk HEA (eds) Endocytobiology Endosymbiosis and Cell Biology, v.1: Walter de Gruyter, Berlin, pp 125–132

Saraswati PK (2008) Life history of foraminifera: stable isotopic and elemental proxies. J Palaeontol Soc Ind 53:1–8

Saraswati PK, Seto K, Nomura R (2004) Oxygen and carbon isotopic variation in co-existing larger foraminifera from Reef flat at Akajima, Okinawa, Japan. Mar Micropaleontol 50:339–349

Segev E, Erez J (2006) Effect of Mg/Ca ratio in seawater on shell composition in shallow benthic foraminifera. Geochem Geophys Geosyst 7:Q02P09. doi:10.1029/2005GC000969

Shackleton NJ (1984) Oxygen isotope evidence for Cenozoic climatic change. In: Brenchley P (ed) Fossils and climate. John Wiley, New York, pp 27–34

Stanley SM (2006) Influence of seawater chemistry on biomineralization throughout Phanerozoic time: Paleontological and experimental evidence. Palaeogeogr Palaeoclimatol Palaeoecol 232:214–236

Stanley SM, Ries JB, Hardie LA (2005) Seawater chemistry, coccolithophore population growth, and the origin of Cretaceous chalk. Geology 33(7):593–596

Urey HC (1947) The thermodynamic properties of isotopic substances. J Chem Soc :562–581

Urey HC, Lowenstam HA, Epstein S, McKinney CR (1951) Measurement of paleotemperatures and temperatures of the Upper Cretaceous of England, Denmark and the southeastern United States. Geol Soc Am Bull 62:399–416

Wefer G, Berger WH (1991) Isotope paleontology: growth and composition of extant calcareous species. Mar Geol 100:207–248

Weiner S, Dove PM (2003) An overview of biomineralization processes and the problem of the vital effect. Rev Mineral Geochem 54:1–29

Zachos JC, Stott LD, Lohmann KC (1994) Evolution of early Cenozoic marine temperatures. Paleoceanography 9(2):353–387

Further Reading

Cuif J, Dauphin Y, Sorauf JE (2011) Biominerals and fossils through time. Cambridge University Press, Cambridge

Norris RD, Corfield RM (1998) Isotope paleobiology and paleoecology. The Paleontological Society Papers, Vol 4. The Paleontological Society: US

Chapter 4
Morphology, Taxonomy and Concepts of Species

4.1 Introduction

The living world displays an amazing range of variation in morphology. The variation is the result of ontogenetic differences (due to age), genetic differences (due to pressure of selection and mutation) and non-genetic differences (due to ecology) among organisms. Fossils show an even higher degree of morphologic variation due to the dimension of time through which the life evolved. The preservation potential of the organism, post-mortem transport, time averaging of assemblages and compaction of sediments are additional factors that introduce morphologic diversity into fossil populations. One cannot predict the amount of variation in a population, but a biologist or paleontologist specializing in a group of organisms gets accustomed to the amount of variation typical of populations within that group and can isolate the unusual variations. Morphology is an essential trait in the classification of the living world. The biological system of classification is referred to as *natural classification*, in which all features are considered for general resemblance rather than particular differences. A biologist or paleontologist classifies the organisms and names them (taxonomy), then determines their evolutionary relationships (phylogeny) and geographic relationships (biogeography). The sciences of taxonomy, phylogeny and biogeography together constitute Systematics. This chapter discusses the method for describing morphology, distinguishing microfossils and inferring phylogenetic relationships through cladistic analysis.

© Springer International Publishing Switzerland 2016
P.K. Saraswati, M.S. Srinivasan, *Micropaleontology*,
DOI 10.1007/978-3-319-14574-7_4

4.2 Quantifying Morphology

Morphology is key to the identification and classification of fossils. Traditionally, morphologic description of fossils has been qualitative to semi-quantitative. Different dimensions of the skeletal elements are measured (as variables) and expressed as statistical parameters of range, mean and variance. Shape of the forms is often expressed as ratios of variables. Good progress has been made over the years in making morphology quantitative so that the taxonomy of fossils is practiced as a rigorous science. Microfossils are particularly amenable to statistical analysis of morphology due to the easy availability of a statistically significant number of well-preserved specimens, compared with fragmentary data in other groups of fossils. The application of quantitative morphology is established well beyond taxonomy to functional morphology and evolution. Due to enhanced computational facilities, quantitative analysis of morphology is becoming routine in taxonomy. The statistical techniques are based on certain assumptions and the users should know if the data to be analysed is suitable for a particular analysis. Some of the methods, for example, require normal distribution and multivariate normality of the data. In using these methods, data is tested for normality and the variables are suitably transformed prior to analysis. The textbooks listed at the end of the chapter discuss the basics of statistics, the assumptions involved and applications of multivariate statistical methods in paleontology.

There are several methods for quantifying morphology, and the choice of method depends on the purpose of the analysis. In most applications, the length, width and height of the shells and dimensions of the chambers are measured and the obtained data is analysed through univariate and multivariate procedures of statistics. Qualitative and multi-state characters can also be treated statistically. Shape is analysed by digitizing the outline of the shells and the coordinates of the equally spaced points are fitted to a Fourier series (see Boon et al. 1982, for stepwise computation, and Muthukrishnan and Saraswati 2001, for application in foraminiferal taxonomy). These two groups of methods are also referred to as conventional morphometrics to differentiate them from the later-developed geometric morphometrics. Geometric morphometrics has gained importance since Bookstein (1991) first presented the methodology and application of this new tool in morphometrics to compare biological shapes. It statistically analyses the coordinates of a set of homologous *landmarks* on the shell. A landmark is a diagnostic point on the shell, for example, an aperture or proloculus in foraminifera. The coordinates of the set of standardized landmarks constitute the Bookstein coordinates of the shape, and these are subjected to multivariate analysis to get information about the shape and shape change.

The measurement of microfossils can be carried out under a stereozoom binocular microscope by micrometre scale or by image analysis software. The preparation of specimens for measurement may be time-consuming. In larger foraminifera, for example, oriented sections are carefully prepared to reveal the internal morphologic details for measurement. The effort, however, is worth taking, considering the quality of information achieved by morphometric analysis. Nowadays, X-ray tomography by micro-CT is enabling three-dimensional measurements of shell morphology without elaborate sample preparation or destruction of the shell (Briguglio et al. 2013).

Theoretical morphology is another way of looking at the shape of the biological forms. It models biological shape to relate form with function. Raup (1966) pioneered this study and constructed a cubic morphospace to represent morphologic variations in mollusks and brachiopods. Theoretical morphology models the existent forms with minimal mathematical complexity by taking a minimum number of parameters. It tries to explain why, in a range of forms produced theoretically, some forms exist in nature while others do not and have never been produced. The analysis is a three-step process:

1. The construction of a theoretical morphospace of hypothetical yet potentially existent morphologies.
2. The examination of the distribution of existent forms in the morphospace to determine which forms are common, rare or non-existent in nature.
3. The functional analysis of both existent and non-existent forms to determine whether the distribution of existent forms is, indeed, of adaptive significance.

Berger (1969) was the first to create a two-dimensional morphospace for planispiral and trochospiral foraminifera, although he did not elaborate the adaptive significance of the generated shapes. Since then, theoretical morphology has received much attention in explaining the functional significance of the gross morphology of foraminiferal shells. Signes et al. (1993) mathematically modelled the growth of planktic foraminifera by establishing four parameters (Box 4.1). They investigated how surface area and shell volume change with growth. Among the four parameters considered in the model, only the proportionality between the consecutive chamber volumes (Kt) is found to influence the values of surface area and volume. It is also noted that both surface area and shell volume change exponentially with size. These observations have important consequences for physiology and the functional requirement of foraminifera because:

1. The increase in outer surface area and consequent increase in porosity may govern the maximal rate of gas exchange, and thus relate to rate of respiration.
2. Shell volume is related to biomass and, therefore, to the amount of oxygen and prey (nutrition) needed.
3. The ratio of total surface area to shell volume is related to the total calcification effort per unit of biomass (energy required for calcification).

The mathematical modelling of shell morphology has given new insights into the growth of shells modulated by the physiological requirements of foraminifera (see Brasier 1982, Renzi 1988 for more examples).

Box 4.1: Mathematical Modelling of Growth of Planktic Foraminifera

Signes et al. (1993) modelled the growth of planktic foraminifera by way of the following four parameters (see Fig. 4.1 for explanation). A number of growth pattern and possible shapes of the shells can be generated through various combinations of these parameters. The functional significance of the generated shapes is discussed in the text.

1. Kt=ratio of chamber volume to the volume of the pre-existing shell ($CVn + 1/SVn$).
2. φ=angle between consecutive chambers (2π/no. of chambers per whorl).
3. Ky=displacement of the chambers along the axis of coiling [$(Yn + 1 - Yn)/(Xn + 1 - Xn)$].
4. D=distance of the centre of the chamber to the axis of coiling divided by the radius of the chamber (Xn/Rn).

Shapes generated at different values of Ky are shown in Fig. 4.1. The side views are shown on the left and the frontal views are on the right. The shell is planispiral at Ky=0 (A); trochospiral at Ky=1 (B); and high trochospiral at Ky=2 (C).

The opening of the umbilicus is achieved by changing the parameter D (Fig. 4.1). The upper row shows spiral views and the lower row shows umbilical views generated at D=0.8 (A, D), D=1 (B, E) and D=1.2 (C, F).

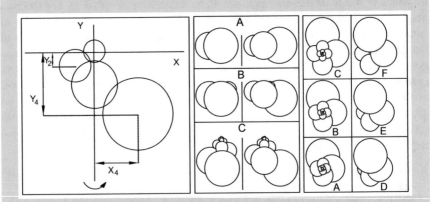

Fig. 4.1 Theoretically simulated growth of planktic foraminifera: the parameters used to generate the shapes (*left*) and shapes generated at different values of Ky (*middle*) and D (*right*) (redrawn after Signes et al. 1993, with permission © the Paleontological Society)

4.3 Taxonomy

The classification and nomenclature of lifeforms constitute taxonomy. Until the 1970s, most paleontologists were *evolutionary taxonomists*. These classical taxonomists used a traditional and flexible combination of criteria to erect a hierarchical classification. Morphological resemblance and phylogenetic relationships were the basis of classification. The order of succession in the rock record and geographical distribution played important parts in establishing phylogenetic relationships. Some taxonomists criticized this method for its uncertainties and subjectivity. A school of *numerical taxonomists* tried to avoid subjectivity by following a quantified phenetic similarity for natural groupings. The basic premises of numerical taxonomy are as follows (Sneath and Sokal 1973):

1. The greater the content of information in the taxa of a classification and the more characters on which it is based, the better a given classification will be.
2. A priori, every character is of equal weight in creating natural taxa.
3. Overall similarity between any two entities is a function of their individual similarities in each of the many characters through which they are being compared.
4. Taxonomy is viewed and practiced as an empirical science.
5. Classifications are based on phenetic similarity.

A large number of characters are chosen in numerical taxonomic classification. The data may be both qualitative and quantitative. The first step in the classification is to estimate similarity among the taxonomic units by calculating the similarity coefficient or distance coefficient. In the next step, the similarity (or distance) matrix is subjected to cluster analysis for hierarchical arrangement of the taxa in the form of a phenogram (Box 4.2). The numerical taxonomy has proven useful in many cases, but subjectivity could not be eliminated because there are a number of algorithms for computing resemblance and there are a number of procedures of cluster analysis. The hierarchical structure of the resulting phenogram can change markedly by altering the procedure.

The biological classification is hierarchic in nature and all natural groups of about the same status are given rank names. Linnaeus was the first to provide a comprehensive scheme, and the rank names used in biological taxonomy derive mainly from him. Linnaeus recognized five ranks, but many additional ranks have come into use over the years. The following example shows the classification of a benthic foraminifer *Nummulites acutus*:

Phylum: Protista Haeckel, 1866.
Class: Rhizopodea von Siebold, 1845.
Order: Foraminiferida Eichwald, 1830.
Suborder: Rotaliina Delage and Herouard, 1896.
Super-family: Nummulitacea de Blainville, 1827.
Family: Nummulitidae de Blainville, 1827.
Genus: *Nummulites* Lamarck, 1801.
Species: *Nummulites acutus* (Sowerby), 1840.

Box 4.2: Numerical Taxonomy

There have been differences of opinion on the validity of two subgenera of
Lepidocyclina, *L.* (*Eulepidina*) and *L.* (*Nephrolepidina*). Numerical taxonomic analysis was carried out on the population of *Lepidocyclina* to see if
the two taxa are statistically valid. A number of variables were measured on
the embryonic apparatus of the genus in equatorial sections (Fig. 4.2). Based
on this, several parameters were calculated (refer to Saraswati 1995 for details
on the calculation of these parameters). In all, seven parameters were used in
analysis; their statistical summary is given in Table 4.1. For numerical taxonomic analysis, taxonomic resemblance is estimated by the Euclidean distance coefficient as follows:

$$d_{ij} = \sqrt{\frac{\sum_{k=1}^{m}\left(X_{ik} - X_{jk}\right)^2}{m}}.$$

Fig. 4.2 Embryonic
apparatus of *Lepidocyclina*

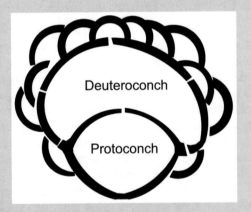

Table 4.1 Morphometric parameters of two subgenera of *Lepidocyclina* used for statistical
analysis

Subgenera		D_I (μm)	D_{II} (μm)	A (%)	X	E (%)	Y	Dc (%)
Eulepidina	Max	1330	1760	86	2.0	100	1.1	127
	Min	480	830	53	1.3	25	0.8	20
	Av	773	1204	73	1.6	82	1.0	78
	SD	181	226	11	0.2	26	0.1	28
Nephrolepidina	Max	400	660	57	2.3	90	1.5	60
	Min	120	200	32	1.2	19	0.8	15
	Av	235	347	42	1.5	34	1.0	27
	SD	72	102	07	0.2	16	0.2	11

(continued)

Box 4.2 (continued)

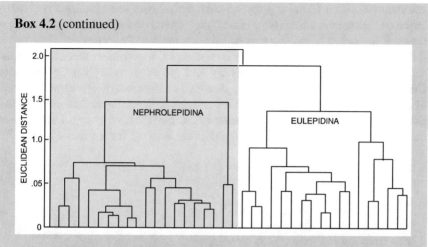

Fig. 4.3 Phenogram illustrating two clusters of taxa corresponding to *Eulepidina* and *Nephrolepidina*

The data matrix of the Euclidean distance coefficient is subjected to cluster analysis. The resulting phenogram (Fig. 4.3) clearly distinguishes two groups. Except for one specimen, all the specimens clustered in two groups correspond to the conventionally identified subgenera.

A classification not only provides information about the morphological attributes of the organism but also reflects its evolutionary relationship with other organisms. There are procedures and rules for the naming of species and the formation of taxonomic categories recommended by the International Code of Zoological Nomenclature (ICZN). While reporting a new species, the researcher should follow the guidelines of the ICZN. In recent years, an important initiative in the field of taxonomy has been taken to prepare global databases of fossil occurrences. The database serves two important purposes. It brings consistency to taxonomy by making the morphological and stratigraphic details of species easily accessible to all and it provides data for addressing large-scale processes in the evolution of life. Some of the databases of particular interest to micropaleontology are discussed later (Sect. 4.6).

4.4   Cladistic Analysis

Both biologists and paleontologists aim to establish the evolutionary or genealogical relationship (phylogeny) of organisms. Biologists use DNA and other molecular data to reconstruct phylogeny. Paleontologists use morphology and the temporal position of the taxa in the geological record to infer ancestral descendant relationships and visually represent them through an *evolutionary tree*. In 1966, a German entomologist, Willi Hennig, proposed cladistic analysis to reconstruct phylogeny based on "shared evolutionary novelties" (or "shared derived characters") and portrayed it through a branching tree called a *cladogram*. Conceptually, cladistics is different from phenetics (or numerical taxonomy). It postulates that classification should only reflect evolutionary history and ignore overall phenetic similarity. Cladistic analysis soon found wide application in paleontology. The central concept of cladistic analysis is that, in any group of organisms, characters are either primitive (plesiomorphy) or derived (apomorphy). Closely related groups have "shared derived characters", called synapomorphies (see Box. 4.3 for terminology).

Box 4.3: The Terminology of Cladistics

Plesiomorphy: a primitive character.

Apomorphy: an advanced or derived character.

Autapomorphy: a derived character shared by a single group.

Synapomorphy: a derived character that is shared by two groups.

Symplesiomorphy: a shared primitive character.

Monophyletic groups contain the common ancestor and all of its descendants
    (D, C, B and A in Fig. 4.4).

Paraphyletic groups are descended from a common ancestor but do not
    include all descendants (B and C, Fig. 4.4).

Polyphyletic groups are the result of convergent evolution. Their representa-
    tives are descended from different ancestors, and hence, although they may
    look superficially similar, any polyphyletic group comprising them is arti-
    ficial (Fig. 4.4).

Fig. 4.4 A cladogram illustrating the different groups of taxa in cladistic analysis

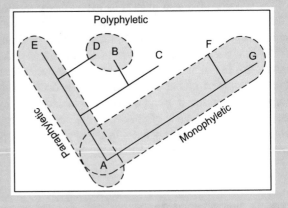

Cladistic analysis begins with construction of a "character matrix", in which rows consist of taxa and columns of characters. The character states in the matrix are generally represented by 0 (for absence) and 1 (for presence), although there are ways to represent multistate characters as well. The next step involves identifying the *polarity* of characters, whether they are primitive or derived. A cladogram is constructed based on shared derived characters. Phylogeny can be complicated and there may be several possible cladograms based on the given data set. The principle of parsimony helps in choosing the best tree. According to this principle, the theory of nature should be the simplest explanation. The cladogram having the fewest steps of character transition for the given data is the most parsimonious tree. A simplified example of cladistic analysis of recent foraminifera belonging to the Soritacea group is discussed in Box 4.4. In practice, however, a much larger data set of species

Box 4.4: Cladistic Analysis of Soritacea

The data matrix showing presence (1) and absence (0) of nine characters in five species of Soritacea is given below. *Peneroplis planatus* is taken as the outgroup.

Character/species	1	2	3	4	5	6	7	8	9
Marginopora kudakajimaensis	1	1	1	1	1	1	1	1	1
Amphisorus hemprichii	1	1	1	1	1	1	1	0	1
Sorites orbiculus	0	1	1	1	0	1	0	0	1
Sorites orbitoides	0	1	1	1	0	1	0	0	0
Sorites bradyi	0	1	0	0	0	1	0	0	0
Peneroplis planatus	0	0	0	0	0	0	0	0	0

The characters are as follows: (1) median apertures between marginal apertural rows; (2) multiple apertural rows of circular or crescentic form; (3) flabelliform chambers; (4) cyclic chambers; (5) A-form embryo possesses a vorhof; (6) internal skeleton consists of septula; (7) duplex skeleton; (8) median skeleton; (9) outer wall with evenly dispersed pits (data after Gudmundsson 1994).

In the following sequence of cladograms (Fig. 4.5), shared derived character (apomorph) determines the classification at each step. The ticked characters on the branches are the apomorphs that define the sister taxa. In step 1, for example, characters 2 and 6 are derived characters that separate the five species from the outgroup. Similarly, other characters are involved in subsequent steps. The final cladogram is highlighted.

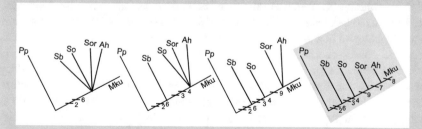

Fig. 4.5 Cladogram illustrating the relationship between the selected species of recent Soritacea

and characters is involved and manual construction may not be possible. Software (Sect. 4.6) is run to generate the most parsimonious or all possible equally parsimonious trees. Cladistic hypotheses have made major advancements since development of the method and readers should refer to the literature for detailed discussions on tree construction (e.g. Eldredge and Cracraft 1980).

4.5 Species Concepts

Species are taxa of the lowest rank in the Linnaean hierarchy of life. It is the smallest unit that a taxonomist identifies in his/her study. Essentially, there are two concepts of species: biological and morphological. Biologists define species as "groups of actually or potentially interbreeding natural populations which are reproductively isolated from other such groups". The application of the above definition relies on the identification of the actual or inferred reproductive potential (interbreeding) between populations. This definition is inappropriate for fossils, because interbreeding in a population cannot be demonstrated and genetic and ecological information other than by inference are lacking. All paleontological species are, therefore, necessarily morphological species, though we can gain some insight into the scales of morphologic variation between species from modern biological species. The paleontological species concept is derived from the "evolutionary species" definition of G. G. Simpson, according to which "species are groups of individuals which exchange genetic information primarily with other members of the group and which share a common evolutionary history". How do species originate? There are several models of speciation (the process by which species are formed). The geographically based models consider physical isolation of populations to be the mechanism of species formation. The ecological models stress the role of differential ecological adaptation and genetic models consider internal genetic mechanism as the cause of species divergence (Lazarus 2003).

How do we identify species? The Linnean system of drawing boundaries between species is based on discontinuities in the range of morphologic variation. At times, it may be arbitrary in the fossil populations that gradually change from one form to another. The difference between the beginning and the end of the gradually evolving population, however, is so great that paleontologists tend to subdivide the lineage, even if it is arbitrary. Taxonomists are often categorized as "splitters" and "lumpers" (Hornibrook 1968). Due to the usefulness of fossils in stratigraphic zonation, micropaleontologists give new names to separable morphological groups even if variation is slight but consistent. On the other side of these splitters are the lumpers, who accept wider variations in species and genera. There are criticisms of excessive splitting of species, but the observation of molecular systematics is interesting in this context. The molecular study of a living foraminifer, *Ammonia*, has shown that there are as many molecular types of the genus as the morphologically distinguished species recorded globally (Hayward et al. 2004). The morphologically separable species are, thus, also distinguishable by molecular type. The identification of a group of specimens as a new species is largely a subjective one, based on the specialist's long experience with the range of variation commonly found in the group of fossils studied and the range found in modern species of related organisms. It is a common understanding that many fish at first glance may look similar, but a fisherman knows how to separate them (for more on species identification and nomenclature, read Prothero 1998).

The skill of a taxonomist to link morphospecies phylogenetically and biochronologically to develop an evolutionary tree is fundamental to the science of biostratigraphy. A recent planktic foraminifer *Neogloboquadrina dutertrei* occurs as two distinct morphogroups in tropical and subtropical areas, respectively. The scanning electron microscopy revealed that the ultrastructures in the two groups are distinctly different. The two groups, however, are linked by intermediate ultrastructures and, thus, represent phenotypic variants within a cline extending from tropical to cool subtropical areas (Srinivasan and Kennett 1976). It required an intensive study of these forms across the latitude and through the Neogene to recognize two evolutionary bioseries in *Neogloboquadrina*, both having been derived from *Globorotalia (Turborotalia) continuosa* (Fig. 4.6). Whether the two morphogroups of *Neogloboquadrina dutertrei* are two distinct "species" or "phenotypic variants" was subject to in-depth observation of evolutionary changes in several morphological traits of the foraminifer.

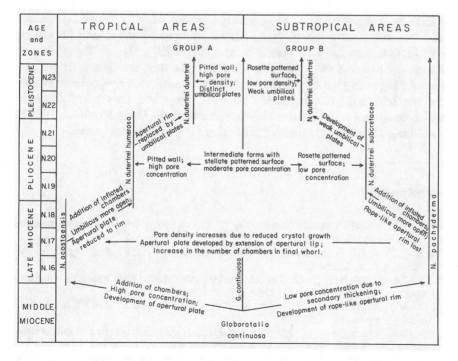

Fig. 4.6 Phenotypic and evolutionary relationships established through comparative morphologic and ultrastructural studies of planktic foraminifera (reproduced after Srinivasan and Kennett 1976)

4.6   Database and Software Program

Database: The taxonomic information, illustrations and stratigraphic distribution of microfossils are available online through several databases. Some of them are listed below.

CHRONOS: Many databases of micropaleontologic interest are available through this portal. Of specific interest in regard to microfossils reported in DSDP and ODP samples are the databases *Neptune* and *Janus*. The *Paleobiology Database* and *PaleoStrat* hosted on this site, despite being of wider use for marine and terrestrial animals and plants, provide taxonomic and stratigraphic information on microfossils.
WoRMS: This is the World Register of Marine Species, a part of which is the World Foraminifera Database cataloguing foraminiferal species.
Palydisk: This is the palynological database of the American Association of Stratigraphic Palynologists.
Nannoware/BugCam: This is managed by the International Nannoplankton Association and has a digital image catalogue of Cenozoic calcareous nannofossils.

Software Programs: The most widely used software for statistical analysis includes SPSS, SAS and Systat. A number of freely downloadable software programs are also available for statistical analysis and morphometric analysis. PAST (Paleontological Statistics) contains a number of statistical methods generally used for paleontological data analysis. Some individual efforts are making computations easily accessible for morphometric analysis. These include *Morphometrics at SUNY Stony Brook* and *Morpho-tools.net*. The latter contains tools for Eigen-shape and landmark analysis and statistical analysis of morphometric data, including linear regression, principal component analysis and canonical variety analysis. Software named Phylogenetic Analysis Using Parsimony (PAUP) and MacClade are used to run cladistic analysis to generate cladograms.

References

Berger WH (1969) Planktonic Foraminifera basic morphology and ecologic implications. J Paleontol 43:1369–1383
Bookstein FL (1991) Morphometric tools for landmark data: geometry and biology. Cambridge University Press, Cambridge
Boon JD, Evans DA, Hennigar HF (1982) Spectral information from Fourier analysis of digitized quartz grain profiles. Math Geol 14:589–605
Brasier MD (1982) Foraminiferid architectural history: a review using the MinLOC and PI methods. J Micropalaeontol 1:95–105
Briguglio A, Hohenegger J, Less G (2013) Paleobiological applications of three dimensional biometry on larger benthic foraminifera: a new route of discoveries. J Foraminifer Res 43:72–87
Eldredge N, Cracraft J (1980) Phylogenetic patterns and the evolutionary processes – method and theory in comparative biology. Columbia University Press, New York

Gudmundsson G (1994) Ontogeny and systematics of Recent Soritacea Ehrenberg 1839 (Foraminiferida). Micropaleontology 40(2):101–155

Hayward B, Holzmann M, Grenfell HR, Pawlowski J, Triggs CM (2004) Morphological distinction of molecular types in *Ammonia* – towards a taxonomic revision of the world's most commonly misidentified foraminifera. Mar Micropaleontol 50:237–271

De B Hornibrook N (1968) A handbook of New Zealand microfossils (Foraminifera and Ostracoda). New Zealand Department of Scientific and Industrial Research, Wellington

Lazarus DB (2003) Species evolution. In: Briggs DEG, Crowther PR (eds) Palaeobiology II. Blackwell, Oxford, UK, pp 133–137

Muthukrishnan S, Saraswati PK (2001) Shape analysis of the nucleoconch of *Lepidocyclina* from Kutch: a taxonomic interpretation. Micropaleontology 47:87–92

Prothero DR (1998) Bringing fossils to life – an introduction to paleobiology. WCB McGraw-Hill, Boston

Raup DM (1966) Geometric analysis of shell coiling: general problems. J Paleontol 40:1178–1190

de Renzi M (1988) Shell coiling in some larger foraminifera: general comments and problems. Paleobiology 14(4):387–400

Saraswati PK (1995) Biometry of early oligocene *Lepidocyclina* from Kutch, India. Mar Micropaleontol 26:303–311

Signes M, Bijma J, Hemelben C, Ott R (1993) A model for planktic foraminiferal shell growth. Paleobiology 19(1):71–91

Sneath PHA, Sokal RR (1973) Numerical taxonomy. W H Freeman, San Francisco, CA

Srinivasan MS, Kennett JP (1976) Evolution and phenotypic variation in the Late Cenozoic *Neogloboquadrina dutertrei plexus*. In: Takayanagi Y, Saito T (eds) Progress in micropaleontology. Micropaleontology Press, American Museum of Natural History, NY, pp 329–355

Further Reading

Drooger CW (1993) Radial Foraminifera: morphometrics and evolution. Verhandelingen der KoninklijkeNederlandseAkademie van Wetenschappen, AfdNatuurkunde, Eerste Reeks, deel 41, North Holland

Hammer O, Harper DAT (2006) Paleontological data analysis. Blackwell, Oxford, UK

Reyment RA (1991) Multidimensional palaeobiology. Pergamon, Oxford, UK

Chapter 5
Basic Concepts of Ecology

5.1 Introduction

Ecology is the interaction of organisms and their environment. The principles of ecology are extended to paleontological records for deciphering the paleoecology of fossil assemblages. Paleoecology is not just the ecology of past organisms but has a wider dimension of examining large-scale and long-term processes beyond the limit of biologists' observations, called "evolutionary paleoecology". Paleoecology is termed *autecology* when it refers to the behaviour of individual organisms and their relation to environment, and *synecology* when the ecology of communities of organisms and their relationship to environment is studied. The fundamental approach to paleoecology is uniformitarian in nature. The ecology of modern organisms and communities is the basis for deciphering the paleoecology of fossil species. It may work in some, while it may not work in many others. The life conditions of many organisms may have changed through geological time and, therefore, the principle of uniformitarianism should not be applied uncritically to fossil records. The paleoecology of Neogene foraminifera, for example, is interpreted with higher confidence than that of the Permian foraminifera. This is because most of the species of the Neogene continue to live or have living relatives with apparently similar tolerance for temperature, salinity and bathymetry. The various principles of physics, chemistry and biology are applied to infer the ecology of the fossil species when the modern analogue is not helpful. There can be various lines of evidence in paleoecology, including genetics, taphonomy, paleobiogeochemistry and stable isotopes.

Most microfossils inhabited a marine environment. Their present-day representatives occur in different parts of the ocean. Some of them are typically marine, such as foraminifera, coccoliths, pteropods, radiolaria and silicoflagellates, and some live in both marine and fresh waters, such as ostracoda and diatoms. Some of the groups are exclusively planktic (floating in water) and others are epifaunal benthic (living on sediments) and infaunal benthic (living in sediments). The oceanic environment

© Springer International Publishing Switzerland 2016
P.K. Saraswati, M.S. Srinivasan, *Micropaleontology*,
DOI 10.1007/978-3-319-14574-7_5

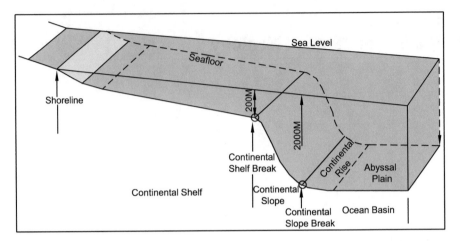

Fig. 5.1 Profile of an ocean floor illustrating continental shelf, continental slope and abyssal plain

is highly diverse in terms of its topography, the temperature and salinity of the water, and ocean currents, as a result of which the ocean provides a diverse and dynamic ecological setting for the life it supports. Geomorphologically, the ocean floor is comprised of four major areas (Fig. 5.1). The *continental shelf* is a broad and nearly flat continental area extending into the sea. At its outer edge, it drops off sharply into the *continental slope*, corresponding to a water depth of ~200 m. The continental shelf is divided into three depth zones: inner shelf (0–30 m), middle shelf (30–100 m) and outer shelf (100–200 m). The continental slope extends to a depth of 2000 m and descends to the *abyssal plain* through a gentler slope of *continental rise*. The broad, flat abyssal plain extends to a depth of about 6000 m. Some oceans have deep, narrow depressions, called trenches, which may extend to more than 10,000 m depth. Life occurs at all depths in the ocean. Living foraminifera are recorded at water depths of >10,000 m in the Mariana trench. It is of ecological significance that the intensity of light and the temperature and salinity of the water change with the depths of the ocean. Three major depth zones in the ocean are recognized, characterized by noticeable changes in the physical parameters. The *surface zone* or *mixed layer* is the warm water layer extending to a depth of 100–500 m, where water is well mixed by wind and waves. A sharp change in temperature, salinity and density of water occurs below the mixed layer, termed the thermocline, the halocline and the pycnocline, respectively. In the *deep zone*, temperature, salinity and density change slowly. It is only the upper 200 m of the seafloor that receives sunlight. Light is a limiting factor in the depth distribution of phytoplanktons and the symbiont-bearing organisms. The other important large-scale ecological factors include ocean currents and upwelling. The ocean current is driven by the salinity and temperature of the ocean water and earth's rotation and wind. Surface currents transport heat from equator to pole and bring cooler water to the equator. Deep cold currents transport oxygen and carry nutrients to shallow water. Some coasts are characterized by *upwelling*, when winds push surface water away from the shore and deep currents of cold water rise to the surface. It brings up nutrients and promotes the growth of plankton.

5.2   Environmental Factors in the Distribution of Organisms

The actual physical environment in which organisms live is the *habitat*. Most habitats are occupied by several species, each with its own *ecologic niche*, defined by the physical, chemical and biological limits of the organisms. The environmental factors have a lower and an upper limit within which a species lives, and these are the maximum stretchable tolerance for the survival of a species. There are, however, critical threshold values beyond which the environment is stressful and an optimum value most favourable for the species at which the abundance is likely to be high. A species may have a larger ecospace in which it can potentially exist, called its fundamental niche, and a smaller part of it, called its realized niche, in which the species actually exists. In an ecosystem of complex biotic and abiotic factors, a species does not have a simple relationship with particular parameters. A number of environmental factors, for example, act in the deep sea to form the foraminiferal assemblage. Some of them may be acting directly and have greater influence than others. In the distribution of benthic foraminifera, temperature and salinity are often considered to be not very important compared with oxygen and organic flux (Van der Zwaan et al. 1999). A model based on infaunal benthic foraminifera from shallow to deep sea environments indicates that varying combinations of organic flux (food) and level of oxygen limit the distribution of foraminifera in sediments (see Box 5.1 for the TROX model of Jorissen et al. 1995). A good understanding of such complexities of the ecological processes is important for developing reliable proxies for paleoenvironmental interpretation.

The main environmental parameters controlling the distribution of organisms are discussed below.

Temperature: Physiology and metabolism of organisms are affected by temperature. The life activities operate best at an optimal temperature. The many organisms that have a limited tolerance for change in temperature are called stenothermic, while others, known as eurythermic, can tolerate a wide range of variation. Stenothermic species are useful in paleotemperature interpretation. In marine environments, temperature variation is low due to the high heat capacity and thermal inertia of seawater. Most marine organisms, therefore, are intolerant of extreme

Box 5.1: TROX Model for Distribution of Foraminifera

Trophic condition and oxygen concentration are of great ecological importance. Jorissen et al. (1995) proposed a model based on these parameters,
popularly known as the TROX model, according to which the distribution of
infaunal benthic foraminifera is limited by the combination of organic flux
and oxygenation (Fig. 5.2). In a eutrophic condition, the redox zone is shallow due to consumption of oxygen in organic matter degradation and, therefore, infauna cannot penetrate deeper. Oxygen, and not food, is the limiting
parameter in the distribution of infauna in this condition. In the oligotrophic
environment of the deep sea, the redox zone resides deeper due to decreased
demand for oxygen. Oxygen is not a limiting parameter for infaunal distribution. In this condition, organic supply is so low that organic matter could be a
limiting parameter. It is, thus, a combination of food and oxygen that influences how deep the infauna penetrates the sediments.

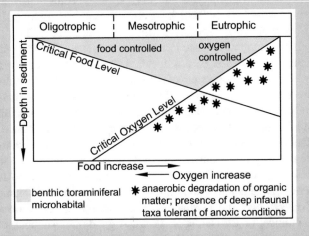

Fig. 5.2 Distribution of infaunal benthic foraminifera in relation to organic flux and oxygenation (redrawn after Jorissen et al. 1995, with permission © Elsevier)

temperatures. The microplankton such as planktic foraminifera live in distinct biogeographic provinces based on their preferred temperature range as follows:

Tropical: 24–30 °C.
Subtropical: 24–18 °C.
Transition: 18–10 °C.
Subpolar: 10–05 °C.
Polar: 0–5 °C.

The seasonal variation in sea-surface temperature is 0–2 °C in the tropics, 5–8 °C in the mid-latitudes and 2–4 °C in the polar regions. Ocean water temperature also varies with depth. The surface layer is warm, the thermocline has a sharp decrease in temperature, and the deep zone has a temperature of <5 °C.

Salinity: This is another important parameter in the distribution of organisms living in water. Some organisms are tolerant of salinity variations (euryhaline) while others have a narrow tolerance for salinity change (stenohaline). Among microfossils, radiolaria are stenohaline but ostracoda, as a group, are euryhaline. Salinity is measured as the total dissolved salts per volume of seawater, commonly expressed as parts per thousand, or per mil (‰). The ranges of salinity for different waters are as follows:

Fresh water	0–0.5‰
Brackish water	0.5–30‰
Normal seawater	30–40‰
Hypersaline water	40–80‰
Brine	>80‰

Oxygen: Oxygen is essential for respiration and, therefore, of critical importance for marine and freshwater organisms. Based on oxygen content, the following levels of oxygenation are defined (Allison et al. 1995):

Oxic	>1.0 ml/l
Dysoxic	1.0–0.2 ml/l
Suboxic	0.2–0 ml/l
Anoxic	0 ml/l

The biofacies terms for different levels of oxygenation are as follows:

Aerobic: normal benthic fauna, no oxygen restriction.
Dysaerobic: impoverished benthic fauna stressed by low bottom water oxygen value.
Anaerobic: no benthic fauna due to lack of oxygen.

The dissolved oxygen level drops rapidly with depth in the ocean. It drops to 1 ml/l at about 400 m, and there is an oxygen-minimum zone between 600 and 1000 m. The low oxygen is caused by heavy respiration by marine life in the near surface waters and due to the decay of most organic matter at this level, depleting the level of oxygen. Below the oxygen-minimum zone, the ocean water is again well oxygenated, because of the flow of oxygen-rich and cold sinking water in Antarctica and the Arctic.

Depth and Light: Depth is an important variable in oceanic environments, because several factors change with depth. Light is an important ecological parameter for sustenance of life. Its intensity in ocean water decreases with depth. The sunlit upper layer of the ocean is the photic zone or epipelagic zone. It extends to a depth of 200 m, but it may be shallower due to the presence of suspended particles in the water or due to high productivity of phytoplankton that may inhibit penetration of light. This is the zone of photosynthesis characterized by calcareous nanno-plankton and symbiont-bearing foraminifera. The other planktic micro-organisms also inhabit this zone. Temperature and dissolved oxygen also decrease with depth, while salinity and carbonate content change with it. Most ocean waters below CCD (the depth at which carbonates dissolve: 3000–4000 m) are under-saturated with calcite, so that organisms with calcareous shells cannot survive and the settling plankton are not preserved.

Nutrients: Nutrient availability is referred to as oligotrophic (low nutrient), mesotrophic (moderate nutrient) and eutrophic (high nutrient). The productivity in the three trophic conditions is as follows:

Oligotrophic	<100 g C_{org} $m^{-2}y^{-1}$
Mesotrophic	100–300 g C_{org} $m^{-2}y^{-1}$
Eutrophic	301–500 g C_{org} $m^{-2}y^{-1}$

The density of phytoplankton increases in a eutrophic condition and it inhibits penetration of light. It becomes a limiting factor in the distribution of symbiont-bearing organisms.

Substrate: The distribution of benthic foraminifera and other benthic microfauna may be restricted by the nature of the substrate on which they live. Some require hard substratum or plants for attachment, while others may require soft substrate to burrow. The feeding mechanism of benthic forms is often related to the type of substratum.

5.3 Quantitative Analysis for Paleoecology

A number of quantitative methods are used to express species richness, diversity and equitability of samples for ecological/paleoecological interpretation. The basic data for computation of these values are species composition and their counts in a sample. For foraminifera, generally, 300 individuals are picked to estimate diversity. The abundance of species may be expressed in an absolute number, as a percentage or as a proportion (=%/100). More often, the sample size (total individuals picked for species count) varies from one study to the next. The species richness is a function of sample size. Comparison of two or more samples of different sizes for their diversity would require standardizing them from the smallest sample size by a *rarefaction curve* (Box 5.2). The rarefied counts are statistically tested for their

Box 5.2: Rarefaction

In picking microfossils from washed residues, one will find that the number of recorded species increases as a greater number of specimens is picked. Estimation of species richness is, thus, biased by sample size. The biased data may lead to incorrect interpretation when samples of different sizes are compared to address the issues of biodiversity change through time or space. The data is, therefore, standardized to remove sample-size bias through a method called rarefaction. The method estimates the number of species that would have been found had a smaller number of specimens been picked from the sample. A rarefaction curve (Fig. 5.3) simply represents the count of species in different sample sizes picked randomly from the entire collection. The expected number of species ($E(S_n)$) in smaller samples is calculated as follows:

$$E(S_n) = \sum_{i=1}^{S} \left[1 - \frac{\binom{N-N_i}{n}}{\binom{N}{n}} \right],$$

where $E(S_n)$ is the estimated number of species for a given number of specimens (n), N is the number of individuals in the original sample, S is the number of species in the original sample, and N_i is the number of individuals of the ith species. Refer to Raup (1975) for further details on the rarefaction curve, computation of $E(S_n)$, and its variance.

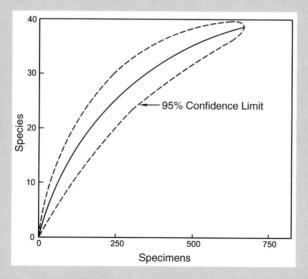

Fig. 5.3 A hypothetical rarefaction curve

Fig. 5.4 Fischer's
α-diversity plot

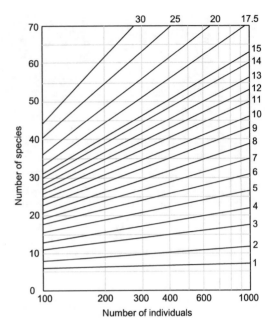

significance. There are various methods of quantitative estimate of diversity. Some of these are discussed below:

The number of species in a sample is known as *Species richness* (S). There are many ways to express biodiversity. Fischer's α diversity is defined as follows:

$$S = \alpha \ln\left(1 + \frac{n}{\alpha}\right).$$

The diversity values may be directly read in Fig. 5.4 for the known values of species richness (S) and number of individuals (N). Another measure of diversity is the Information Index (H), also known as the Shannon–Weiner index, and expressed as

$$H = -\sum_{i=1}^{n} p_i \ln(p_i),$$

where p_i is the proportion of the ith species. H will attain maximum value if all species in a sample are equally abundant and $H_{max} = \ln S$. There are also indices of evenness (or equitability) that allow us to know the distribution of individuals among the species. The Pielou Equitability Index, varying from 0 to 1, is expressed as

$$J = \frac{H}{H_{max}} = \frac{H}{\ln S}.$$

Evenness is also expressed as

$$E = \frac{e^{H}}{S} \text{ (where, } e=2.718).$$

The S, H and E are related through the following linear relationship (Buzas and Hayek 1998), often referred to as SHE analysis:

$$H = \ln S + \ln E.$$

SHE analysis has found applications in biofacies analysis and biozonation.

No single index adequately describes the faunal diversity and, therefore, use of more than one index of diversity is recommended for more accurate results. If the assemblage contains more rare species, each represented by a few individuals, the species richness (S) is prone to the sample size. The information function or equitability index considers both numbers of species and their abundance.

Multiple variables are involved in paleoecological studies. They may contain a number of measured or inferred ecological variables in addition to the taxonomic assemblages. One or more of the several multivariate statistical procedures analyse such data sets. Essentially, these procedures are for understanding the similarity in taxonomic assemblages of two or more samples, to explore groups of ecologically comparable taxa and to examine trends in such groupings. The cluster analysis is a routine procedure for exploring natural groupings of species or samples. Principal component analysis and correspondence analysis provide means to see similarities and differences in samples on comprehensible plots. Hammer and Harper (2005) give the details about the kind of data, data transformation and suitability of the method. Parker and Arnold (2003) have reviewed the statistical methods for the ecology of foraminifera. The statistical methods are tools for understanding the structure of the data, and certainly provide an objective insight into the problem, but the user of the results should have a good understanding of the ecological processes behind the statistically determined species assemblages. Micropaleontologists have developed "transfer functions" to estimate sea-surface temperatures and sea level changes in the Quaternary (Box 5.3), and these have contributed significantly to quantifying paleoclimate.

Box 5.3: Transfer Function

The ecological data are statistically analysed to reconstruct paleoenvironment and paleoclimate. Transfer functions are equations derived from statistical analysis of species and the corresponding environmental variables.
The general relationship between microfaunal composition and the corresponding environmental conditions in core top samples, expressed by the
transfer function, is applied to census data of microfossils from core samples to infer the past environmental conditions (see Guiot and de Vernal
2007 for more details). Foraminifera, diatoms, coccoliths, dinoflagellates
and pollens have been the main biological proxies in developing transfer
functions. The estimation of sea-surface temperatures by transfer function
is explained in the flow chart (Fig. 5.5). The commonly used statistical
techniques include principal component regression, partial least squares
analysis, canonical correlation analysis and correspondence analysis.
Imbrie and Kipp (1971) were the pioneers in developing this technique for
paleoclimate research. They proposed the following equation (transfer
function) to represent summer surface-water temperatures (T) as a function
of planktic foraminiferal assemblages:

$$T\left(^{\circ}C\right) = 19.7A + 11.6B + 2.7C + 0.3D + 7.6,$$

where A to D represent the values of tropical (A), subtropical (B), subpolar
(C) and gyre margin (water mass peripheral) (D) assemblages. The transfer
function is applied to core samples and has given robust interpretation of
sea-surface temperatures during the last glacial maximum at 18 ka.

Foraminifera-based transfer functions have also been developed to
reconstruct sea level changes in the Holocene. The modern-day marsh foraminifera comprise species of *Jadammina*, *Trochammina*, *Ammonia*,
Haynesina, *Elphidium* and *Quinqueloculina* along the high marsh to tidal flat.

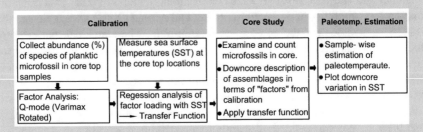

Fig. 5.5 A flow chart to explain the derivation of transfer function for estimating sea-surface temperatures based on microfossils

(continued)

> **Box 5.3** (continued)
>
> It is statistically determined if elevation (calculated as a standardized water level index) is a significant controlling factor for the foraminiferal assemblages in the modern training set. The transfer function is then applied to calibrate fossil assemblages in the core or outcrop samples to get the elevation of each sample with respect to the past mean tide level (see Horton and Edwards 2005 and references therein for methodology).

References

Allison PA, Wignall PB, Brett CE (1995) Paleo-oxygenation: effects and recognition. In: Bosence, DWJ, Allison, PA (eds) Marine palaeoenvironmental analysis from fossils. Geological Society London Special Publications. 83: 97–112

Buzas MA, Hayek LC (1998) SHE analysis for biofacies identification. J Foraminifer Res 28:233–239

Guiot J, de Vernal A (2007) Transfer functions: methods for quantitative paleoceanography based on microfossils. In: Hillaire-Marcel C, de Vernal A (eds) Proxies in Late Cenozoic Paleoceanography. Elsevier, Amsterdam

Hammer O, Harper AT (2005a) Paleontological data analysis. Wiley-Blackwell, Chichester, UK

Horton BP, Edwards RJ (2005) The application of local and regional transfer functions to the reconstruction of Holocene sea levels, north Norfolk, England. The Holocene 15:216–228

Imbrie J, Kipp N (1971) A new micropaleontological method for quantitative paleoclimatology: application to a late Pleistocene Caribbean core. In: Turekian KK (ed) The Late Cenozoic glacial ages. Yale University Press, New Haven, CT, pp 71–181

Jorissen FJ, de Stigter HC, Widmark JGV (1995) A conceptual model explaining benthic foraminiferal microhabitats. Mar Micropaleontol 26:3–15

Parker WC, Arnold AJ (2003) Quantitative methods of data analysis in foraminiferal ecology. In: Sen Gupta BK (ed) Modern foraminifera. Kluwer Academic, New York, pp 71–92

Raup DM (1975) Taxonomic diversity estimation using rarefaction. Paleobiology 1(4):333–342

Van der Zwaan GJ, Duijnstee IAP, den Dulk M, Ernst SR, Jannink NT, Kouwenhoven TJ (1999) Benthic foraminifers: proxies or problems? A review of paleoecological concepts. Earth Sci Rev 46:213–236

Further Reading

Gooday AJ (2003) Benthic foraminifera (Protista) as tools in deep water palaeoceanography: environmental influences on faunal characteristics. In: Southard AJ (ed) Advances in marine biology. Elsevier, Oxford, UK

Hammer O, Harper AT (2005b) Paleontological data analysis. Wiley-Blackwell, Chichester, UK

Murray JW (2006) Ecology and applications of benthic foraminifera. Cambridge University Press, Cambridge, UK

Part II
An Overview of Microfossils

Chapter 6
Calcareous-Walled Microfossils

6.1 Foraminifera

Foraminifera have the distinction of being the most important group of microfossils due to their stratigraphic significance and their value as indicators of paleoenvironment. They occur in a wide range of environments, from marginal marine to deep marine at present, and have been recorded in rocks of such similar environmental settings as those of the Cambrian and younger ages. Some of the species occur in such great numbers that their accumulation on the deep-sea floor constitutes an important lithological unit called foraminiferal ooze. The oozes formed beyond the continental slope are composed primarily of planktic foraminifera. The benthic foraminifera, similarly, are a major constituent of shallow-water carbonates, including the well-known Fusulinid limestone and Nummulitic limestone of the Permo-carboniferous and Eocene ages, respectively. Foraminifera are unicellular animals, similar to Amoebae, and, therefore, belong to the Protista kingdom, but differ from other protists in having thread-like, anastomosing pseudopodia (granuloreticulopodia) and in possessing a shell or test. The molecular data, however, suggests that there are some freshwater foraminifera that do not have tests. Foraminifera have adopted both benthic and planktic modes of life. They generally range in size from 0.1 to 1.0 mm, but some representatives attain the gigantic size of more than 10 cm. Their tests may be non-mineralized, formed from foreign particles cemented together or composed of calcium carbonate and, in rare cases, silica. About 2200 species of extant benthic foraminifera and about 40 species of planktic foraminifera have been reported and described. The total number of living species of benthic foraminifera is estimated to range from 3200 to 4200 (Murray 2007).

© Springer International Publishing Switzerland 2016
P.K. Saraswati, M.S. Srinivasan, *Micropaleontology*,
DOI 10.1007/978-3-319-14574-7_6

Biology of Foraminifera

The cell structure and organelles of foraminifera are like those of the other single-celled organisms. The possession of granuloreticulopods (Fig. 6.1), however, distinguishes foraminifera from the other protists. The cell includes a Golgi body (for secretion of digestive enzymes and other macromolecules), mitochondria (for respiration), ribosome (for protein synthesis) and vacuoles (filled with organic or mineral matter). The cell may be uninucleate or multinucleate. The cytoplasm of foraminifera is partly enclosed within the test and partly forms a thin layer around it. The inner layer of the cytoplasm is the endoplasm and the outer layer is called the ectoplasm. The endoplasm of certain benthic and planktic foraminifera contains diatom, dinoflagellate or other algal symbionts. A thread-like branching network of granular and reticulose pseudopodia extrudes through one or several openings (aperture) in the test. The bidirectional streaming pseudopodia perform several functions, including attachment and locomotion, capturing food, digestion and test construction. The benthic foraminifera move along sediments, seagrass or algal fronds on their pseudopodia. The streaky surface of pseudopodia absorbs food and streams it into the test. The waste products in the test are agglomerated into smaller particles and streamed out of the cell by the pseudopodia.

Fig. 6.1 Schematic diagram to illustrate the cell structure of a foraminifer

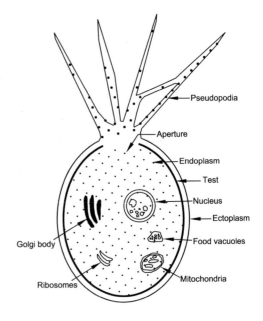

Foraminifera get nutrition from a wide variety of sources. Many of them feed on algae and bacteria. Some of the foraminifera take dissolved organic matter through cell surfaces and some are suspension feeders. Planktic foraminifera ingest diatoms and other algae, silicoflagellates and copepods. Deep-water foraminifera feed upon phytodetritus from the plankton in the surface waters that settles on the seafloor. The degraded organic detritus and bacteria are important food resources for shallow to deeper infaunal taxa (Goldstein 1999). A number of benthic and planktic foraminifera obtain energy through algal symbiosis. They harbour diatoms, dinoflagellates, chlorophyte and rhodophyte. The symbionts utilize P, N and respiratory CO_2 released by the host foraminifera and provide photosynthates and O_2 to the host. Foraminifera modify their tests for the benefit of algal symbionts. Thinning of the test wall, a pitted surface and surface pustules are regarded as morphological adaptations by larger symbiont-bearing benthic foraminifera. In planktic foraminifera, the symbionts are expelled outside on the spines in the day time and retracted inside in the evening.

Foraminifera have a heterophasic life cycle comprising alternation of asexual (schizogony) and sexual (gamogony) phases of reproduction (Fig. 6.2). The heterophasic life cycle results in dimorphism in foraminifera. The individuals reproduced by schizogony (the gamonts) have a large proloculus (the initial chamber) and smaller test compared with those produced by gamogony (the agamonts, also called the microspheric schizonts). The tests with large proloculus and of small size are megalospheric (A-form) and those with small proloculus and of large size are microspheric (B-form). A third generation occurs in some species, the megalospheric schizonts. In view of this, the three generations are preferably called

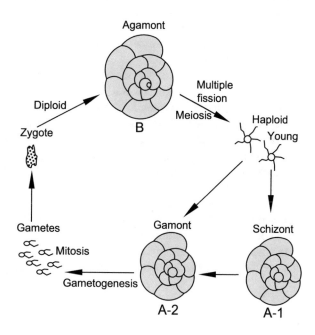

Fig. 6.2 Illustration of life cycle of foraminifera showing alternation of generations (redrawn after Goldstein 1999, with permission © Springer Science + Business Media)

gamont, agamont and schizont, and the corresponding tests are A_2, B and A_1, respectively (Röttger 1992). The initial protoplasm in agamonts is small, but gamonts and schizonts receive large masses of protoplasm due to cell division. It is also significant that, in asexual reproduction (in larger benthic foraminifera), the symbionts are transferred to offspring during fission.

Test Morphology

The morphology of foraminifera is highly variable, from a simple single-chambered tube to a multi-chambered and extremely complex test with canaliculated internal structures. The diverse morphology, however, is functionally advantageous to the unicellular organism in adapting to different lifestyles. The simple tubular morphology of a suspension-feeding epifaunal foraminifer to canaliculated internal structures and high surface area for algal photosymbionts is adaptive. The major morphological variations are in (1) wall composition and ultrastructures, (2) chambers, (3) aperture, (4) accessory structures, and (5) ornamentation.

Wall Composition and Ultrastructures

The test of a foraminifer is composed of organic matter, agglutinated particles or secreted minerals. The organic-walled tests are thin and flexible, made of proteinaceous or pseudochitinous matter, known as tectin. The agglutinated tests are formed through cementing of assorted particles from the seafloor. Foraminifera selectively collect mineral or skeletal particles and bind them together with organic, calcareous or ferric oxide cement. The biomineralized secreted tests are composed of calcite, aragonite or silica.

The secreted tests are further distinguished as microgranular, porcelaneous and hyaline types. The microgranular wall is formed of closely packed microgranules of calcite without detectable cement. The microgranules may be randomly oriented or aligned normal to the surface of the test, giving a fibrous appearance to the wall. The two orientations may occur together as a multi-layered wall. The thin section of a microgranular wall appears dark in transmitted light and opaque when viewed in reflected light. The porcelaneous wall consists of randomly arranged tiny crystals of high magnesium calcite with thin inner and outer veneers of horizontally arranged calcite crystals (Fig. 6.3). It lacks mural pores and appears white and porcelain-like in reflected light and of an amber colour in transmitted light. The hyaline wall is comprised of calcite or aragonite crystals whose c-axes are arranged perpendicular to the test surface. It is perforate, and under the crossed nicols of an optical microscope, it is characterized by a black cross polarization figure with concentric rings of colour.

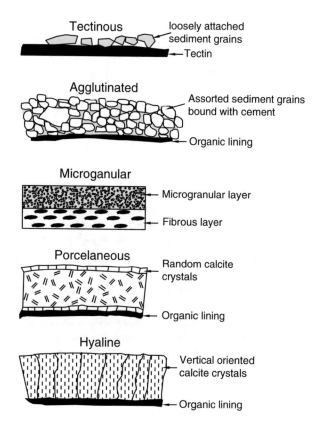

Fig. 6.3 Wall structures in foraminifera

Some of the hyaline foraminifera are optically granular and show a multitude of tiny flecks of colour, but no polarization figure when viewed under crossed nicols. The hyaline foraminifera construct their tests by precipitating calcite crystals on a tectinous template, unlike porcelaneous foraminifera that secrete tiny crystals of calcite within cytoplasmic vesicles and export outside to construct the test (see Sect. 3.3). The wall composition evolved successively from organic to agglutinated and secreted calcite or aragonite types. Among the secreted tests, the early walls with randomly arranged crystals of high-Mg calcite evolved into walls with equidimensional or microgranular crystals of random orientation and then to a preferred orientation of low-Mg calcite or aragonite crystals.

The wall of a foraminifera test has non-lamellar or lamellar ultrastructures. In the non-lamellar type, there is no overlap of previous chamber walls by the new wall, but in the lamellar type, as a new chamber is added, a layer of shell material is also secreted over the exposed earlier part of the test. The lamellar walls are of monolamellar, bilamellar or rotaliid types. In the rotaliid type, the inner lamella also coats the previous apertural face (Fig. 6.4).

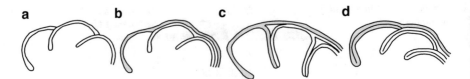

Fig. 6.4 Non-lamellar (**a**), mono-lamellar (**b**), rotaliid (**c**) and bilamellar (**d**) walls in foraminifera

Chamber

The tests of foraminifera consist of either single chambers or multiple chambers separated by septa. The chambers vary widely in shape. They can be tubular, sub-spherical, cylindrical, angular conical, radial elongate, fistulose or cuneate. In certain benthic foraminifera, the chambers in the equatorial plane are subdivided into chamberlets of various shapes, including rectangular, arcuate and hexagonal. Chambers are added in different patterns as the foraminifera grow (Fig. 6.5). They are arranged in a series (uniserial), coiled along an axis in a plane (planispiral), coiled in different planes up the axis of coiling (trochospiral) or coiled in different planes with a shifted axis of coiling (streptospiral). In certain groups, the test is characterized by a continuous spiral with successively formed chambers placed in different planes at 144°, 120° or 180°, known as the milioline type. In coiled tests, the side showing the traces of whorls is termed the spiral side and the opposite side is the umbilical side. Suture, the junction of the septum with the test wall, is flushed, raised or depressed.

Aperture

The chambers in multi-chambered foraminifera are interconnected by *foramen* for the flow of cytoplasm and an aperture in the final chamber to connect it with the exterior. The number, shape and position of the aperture vary considerably (Fig. 6.6). The primary aperture is the main opening(s) in unilocular tests or the final chambers of multilocular tests. It may be single or multiple, rounded, slit-like, loop-shaped or cribrate, among other such variations. The apertures are located at the open end of the tubular forms, at the base of the apertural face (interiomarginal), terminal, peripheral and umbilical. The supplementary aperture occurs either in the apertural face or in the sutural position, in addition to the primary aperture (Fig. 6.7). The accessory aperture does not lead directly to the chamber cavity, but extends beneath the accessory structures. Several foraminifera are characterized by canals that facilitate the cytoplasm deep inside the early parts of the test in connecting directly with the chambers of successive whorls and with the exterior. The canals on the periphery are characterized by numerous anastomosing grooves and ridges and form the marginal cord.

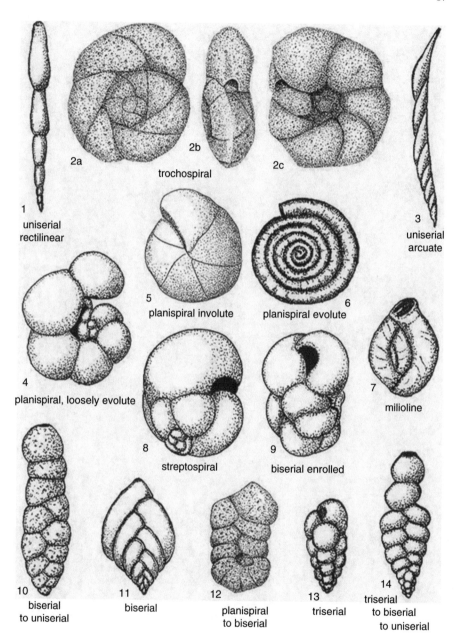

Fig. 6.5 Types of chamber arrangement in foraminifera (from Treatise on Invertebrate Paleontology, Part C, Protista 2, Volume 1, courtesy of and ©1964, The Geological Society of America and the University of Kansas)

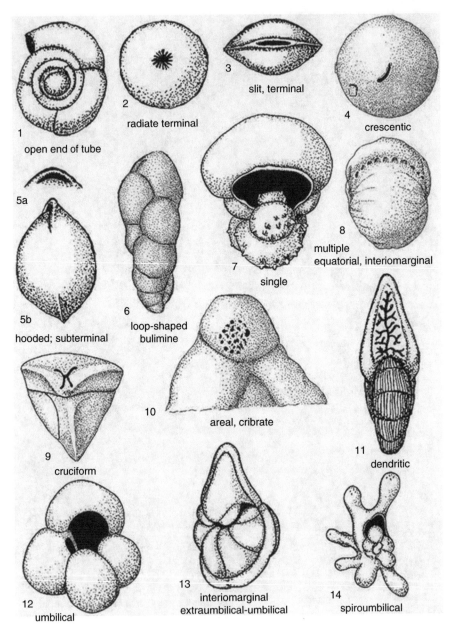

Fig. 6.6 Types of primary aperture in foraminifera (from Treatise on Invertebrate Paleontology, Part C, Protista 2, Volume 1, courtesy of and ©1964, The Geological Society of America and the University of Kansas)

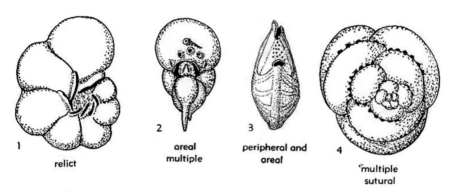

Fig. 6.7 Types of supplementary apertures in foraminifera (from Treatise on Invertebrate Paleontology, Part C, Protista 2, Volume 1, courtesy of and ©1964, The Geological Society of America and the University of Kansas)

Accessory Structures

The accessory structures modify the aperture and comprise the apertural tooth, umbilical teeth, flap, flange, bulla, tegilla and the phialine lip (Fig. 6.8).

Ornamentation

The surface of foraminifera may be smooth or characterized by different types of ornamentation, including spines, nodes, keels, pillars, striations and ribs. The hyaline and porcelaneous foraminifera are ornamented, but this condition is little visible in agglutinated types. These are highly variable, and change within a species from one environment to the other.

Classification

The classification of foraminifera has evolved continually since d'Orbigny (1826) coined the name Foraminiferes and recognized them as a separate order within the Cephalopoda class. The presence of an internal shell, lack of siphon, closed final chamber and apertures for communication between the chambers were regarded as distinguishing characteristics separating them from cephalopoda. D'Orbigny (op cit) further classified foraminifera into several families based on chamber arrangement. The wall composition and structure, chamber arrangement and aperture character have remained the basis for all the later classifications, albeit with different weightage by different workers. The understanding of evolutionary trends in the morphology of foraminifera and the applications of transmission and scanning electron microscopy

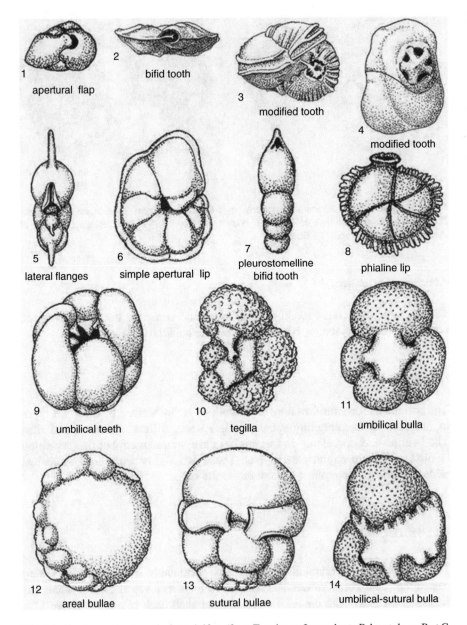

Fig. 6.8 Accessory structures in foraminifera (from Treatise on Invertebrate Paleontology, Part C, Protista 2, Volume 1, courtesy of and ©1964, The Geological Society of America and the University of Kansas)

in revealing surface ultrastructures contributed significantly to updating the classification of foraminifera. The classification proposed by Loeblich and Tappan (1964) has been the most acceptable and most widely followed in the past 50 years. In this classification, foraminifera were accorded the status of Order (Foraminiferida) and the wall composition and texture formed the basis for separation of five suborders. The mode of chamber addition and lamellar characters (in hyaline forms) and unilocular or multilocular tests (in agglutinate and microgranular forms) were used to define superfamilies. The suborders included Allogromiina (organic), Textulariina (agglutinated), Fusulinina (microgranular), Miliolina (porcelaneous) and Rotaliina (hyaline). Later, Loeblich and Tappan (1987, 1992) revised the earlier classification and raised the rank of foraminifera from order to class, referring to them as Class Foraminiferea. In the revised classification, the number of orders increased to 14. Sen Gupta (1999) modified it to retain the identity of Silicoloculinida and Involutinida, also retaining the class name Foraminifera (and *not* Foraminiferea), and provided a simplified key to the identification of the orders. Kaminski (2004) suggested a few modifications in the agglutinated group: suppression of the order Trochamminida, inclusion of Carterinida within the Trochamminacea, and use of a new order Loftusiida. The simplified identification key for foraminiferal orders (Sen Gupta op cit) with amendments for the agglutinated group is presented in Fig. 6.9 and their details are given in Table 6.1.

Molecular Systematics

The molecular biology of foraminifera has provided important insight into their systematics. Notwithstanding the limitation of the method for fossil foraminifera, the genomic data on modern foraminifera complement the conventional morphology-based taxonomy and phylogeny. Some of the important conclusions of molecular studies in reference to the taxonomy of foraminifera are as follows:

1. The molecular data suggests that the freshwater protist *Reticulomyxa filosa* (without a test) is actually a foraminifer. The definition of Foraminifera is, thus, broadened to include all granuloreticuloseans, regardless of their habitat and presence/absence of a test (Pawlowski and Holzmann 2002).
2. The updated SSU rDNA phylogeny of foraminifera has produced a new supraordinal classification of foraminifera (Pawlowski et al. 2013). According to this, the rank of foraminifera is raised to Phylum. Three main groups are recognized: Class Globothalamea (multi-chambered, globular chambers), Class Tubothalamea (multi-chambered, tubular chambers) and a paraphyletic assemblage of "monothalamids" (single-chambered, organic and agglutinated wall).
3. Based on phylogenetic analysis of partial LSU rDNA sequences obtained from living *Ammonia* specimens and their morphometric analysis, it is observed that each molecular type can be distinguished morphologically and can be regarded as a separate species (Hayward et al. 2004). Thus, morphologically separable species are also genetically distinct.

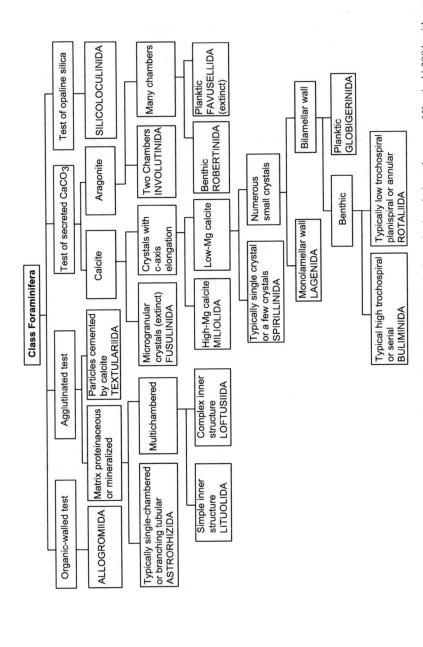

Fig. 6.9 Key to classification of foraminifera (modified after Sen Gupta 1999 to include the revised agglutinated groups of Kaminski 2004, with permission © Springer Science + Business Media)

Table 6.1 Brief description of foraminiferal orders (after Sen Gupta 1999; Kaminski 2004)

Wall type	Order	Description	Example
Organic	Allogromiida	Organic-walled, sometime with few particles attached to it; single-chambered	Allogromia
Agglutinated	Astrorhizida	Matrix proteinaceous or mineralized; single-chambered; tubular or branching	Bathysiphon, Rhabdammina
	Lituolida	Matrix proteinaceous or mineralized; multi-chambered; simple inner structure; usually planispiral, but streptospiral or trochospiral in some cases, uniserial, biserial or multiserial	Miliammina, Haplophragmoides
	Loftusiida	Matrix proteinaceous or mineralized; multi-chambered; complex inner structure—alveolar wall	Cyclammina, Orbitolina
	Textulariida	Particles cemented by calcite; trochospiral, triserial, biserial; chambers simple or may have internal partitions or pillars	Textularia. Eggerella
Opaline silica	Silicoloculinida	Multi-chambered, coiled; imperforate; trochospiral	Miliammellus
Secreted wall-calcite	Fusulinida	Microgranular crystals; extinct	Fusulina, Neoschwagerina
	Miliolida	Crystals with c-axis elongation; high-Mg calcite; imperforate	Quinqueloculina, Alveolinella
	Spirillinida	Crystals with c-axis elongation; low-Mg calcite; optically a single crystal, a few crystals or a mosaic of crystals; planispiral to trochospiral	Spirillina, Patellina
	Lagenida	Crystals with c-axis elongation; low-Mg calcite; mono-lamellar wall; single-chambered or multi-chambered, serial or planispiral	Nodosaria, Lagena
	Buliminida	Crystals with c-axis elongation; low-Mg calcite; bilamellar wall; biserial, triserial or uniserial; aperture with tooth plate in some forms	Bulimina, Bolivina
	Rotaliida	Crystals with c-axis elongation; low-Mg calcite; bilamellar wall; multilocular; planispiral to trochospiral, or may be biserial to uniserial; aperture simple or with internal tooth plate; canal system may be present	Ammonia, Nummulites
	Globigerinida	Crystals with c-axis elongation; low-Mg calcite; bilamellar wall; planktic; perforate	Globigerina, Orbulina
Secreted wall-Aragonite	Involutinida	Two-chambered	Planispirillina
	Robertinida	Planispiral to trochospiral; chambers with internal partitions	Hoeglundina
	Favusellida	Trochospiral with globular chambers; aperture umbilical to extra-umbilical; planktic (extinct)	Favusella

Ecology

Most foraminifera occur in marine environments, but some also occur in estuarine and fresh waters. As a group, they exhibit broad ecological tolerance to salinity, depths and temperature of the ambient waters. The individual species (or genera), however, often have narrow ecological distribution, and this combined attribute makes them highly useful in paleoenvironmental interpretation.

The organic-walled Allogromiid foraminifera occur in both freshwater and marine environments. The agglutinated foraminifera are found in marsh to deep marine environments and with exclusive occurrence below the CCD (ca ~4000 m), due to the absence of calcareous foraminifera. The deepest marine foraminifera, recorded at depths below 10,000 m in the Pacific, are agglutinate taxa comprising the species *Hormosina*, *Reophax* and *Rhabdammina*. The agglutinated taxa belonging to orders Lituolida and Astrorhizida are found in bathyal and abyssal environments. The common genera include *Ammobaculites*, *Ammodiscus*, *Astrorhiza*, *Cyclammina*, *Eggerella* and *Saccammina*. Marsh and mangroves are the other environments preferred by agglutinated taxa. The characteristic assemblage includes *Miliammina*, *Jadammina*, *Arenoparella*, *Haplophragmoides* and *Trochammina* (Fig. 6.10).

The porcelaneous foraminifera belonging to the order Miliolida occur from marginal marine to deep marine environments and tolerate hypersaline or hyposaline conditions. The maximum diversity of miliolid foraminifera is on the inner shelf. The foraminifera belonging to other calcareous orders, including Nodosariida, Buliminida, Robertinida and Rotaliida, are also distributed at all depths of ocean, from shallow marine to deep marine. There are, however, distinct assemblages of foraminifera corresponding to shallow waters (Fig. 6.10) and deep waters (Fig. 6.11), respectively. The details of the depthwise assemblages are discussed further in Sect. 11.2. The distribution of modern deep-water benthic foraminifera appears to be controlled more by bottom water masses (Streeter 1973), trophic conditions and oxygen concentration (Jorissen et al. 1995) rather than by bathymetry alone. The data challenges the validity of some paleoecologic concepts based on empiric observations that are employed for quantitative assessment of paleodepth.

The larger benthic foraminifera belonging to the Miliolida and Rotaliida are shallow marine dwellers and prefer carbonate platforms and reef environments. These foraminifera harbour algal symbionts and, therefore, require sun-lit water, free of turbidity. The reef flats are characterized by high abundance of *Peneroplis*, *Marginopora*, *Amphisorus*, *Sorites* and *Calcarina*. Some of the reef flat taxa continue into the fore-reef, but the typical genera include *Dendritina*, *Alveolinella*, *Parasorites*, *Heterostegina* and *Cycloclypeus* (Fig. 6.12). The distribution of larger benthic foraminifera is also limited by latitude. They occur in low latitudes and require minimum sea surface temperatures of 18 °C.

The planktic foraminifera have a bipolar biogeographic distribution. Five major provinces are recognized: polar, subpolar, transition (temperate), subtropical and tropical. The latitude-limited distribution of planktic foraminifera

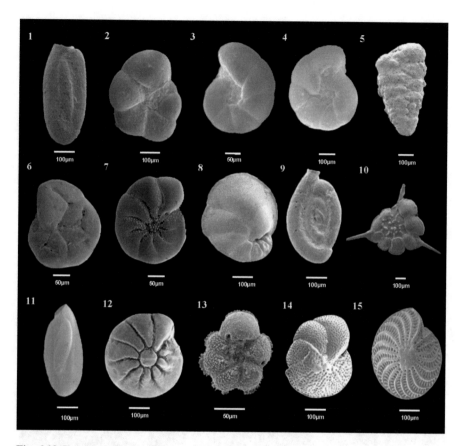

Fig. 6.10 Representative benthic foraminifera from marginal marine and shallow marine environments: *1. Miliammina fusca, 2. Haplophragmoides wilberti, 3. Trochammina inflata, 4. Jadammina macrescens, 5. Textularia sp., 6. Rosalina bradyi, 7. Haynesina germanica, 8. Cancris sp., 9. Spiroloculina sp., 10. Asterorotalia trispinosa, 11. Quinqueloculina ankeriana, 12. Ammonia beccarii, 13. Murrayinella minuta, 14. Cibicides sp., 15. Elphidium crispum* (courtesy Anupam Ghosh)

(Fig. 6.13) suggests that temperature is an important control in their distribution. The other factors that determine the planktic foraminiferal assemblages in modern oceans include salinity, nutrient, turbidity, water illumination and the hydrodynamics of water masses. Although planktic foraminifera generally occur away from the shore in a normal marine, open ocean environment, in laboratory culture, *Globigerinoides ruber* is found to sustain a much wider salinity range of 22–49‰ (Bijma et al. 1990). In the same experiment, it was also noted that the temperature tolerances of different species compare well with their global temperature distribution patterns. The triserial planktic foraminifera, including *Galitellia* and *Guembelitria*, are known to survive in environmentally unstable conditions of shelf and marginal seas and upwelling areas (Kroon and Nederbragt 1990). The

Fig. 6.11 Representative deep-water benthic foraminifera: (agglutinated) *1. Textularia lythostrata* X 39; *2. Siphotextularia finlayi* X 81; *3. Karreriella siphonella* X 44; *4. Eggerella bradyi* X 66; (porcellaneous) *5. Quinqueloculina seminulum* X 458; *6. Biloculina murrhina* X 503; (hyaline) *7. Ceratobulimina pacifica*; *8. Bolivinita quadrilatera* X 71; *9. Brizalina alata* X101; *10. Bulimina striata*; *11. Uvigerina peregrina*; *12. Neouvigerina proboscidea* X89; *13. Cibicides robertsonianus* X98; *14. Oridorsalis*; *15. Pullenia bulloides* X 547; *16. Melonis pompilioides* X101; *17. Melonis pompilioides* X67; *18. Osangularia bengalensis* X77

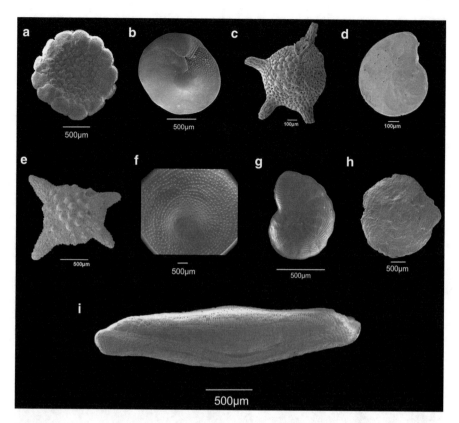

Fig. 6.12 Some representative genera of recent larger benthic foraminifera: *Planorbulina* (**a**), *Amphistegina* (**b**), *Calcarina* (**c**), *Heterostegina* (**d**), *Baculogypsinoides* (**e**), *Cycloclypeus* (**f**), *Peneroplis* (**g**), *Sorites* (**h**), *Alveolinella* (**i**)

planktic foraminifera float in the water column and occupy a range of depths. The species of *Globigerinoides* inhabits shallow waters. *Globigerina bulloides*, *Hastigerina pelagica* and *Orbulina universa* are some of the intermediate water species living in 50–100 m depths. The deeper species living below 100 m include *Globorotalia menardii*, *Sphaeroidinella dehiscens* and *Neogloboquadrina pachyderma*, among others. It is noteworthy that planktic foraminifera may change the depth of their habitat diurnally, seasonally or ontogenetically.

Geologic History

The agglutinated foraminifera *Ammodiscus*, *Glomospira*, *Turritellella*, *Platysolenites* and *Spirosolenites* are the earliest genera recorded from Early Cambrian, shallow marine, siliciclastic shelf deposits (Culver 1991, McIlroy et al. 2001). These are

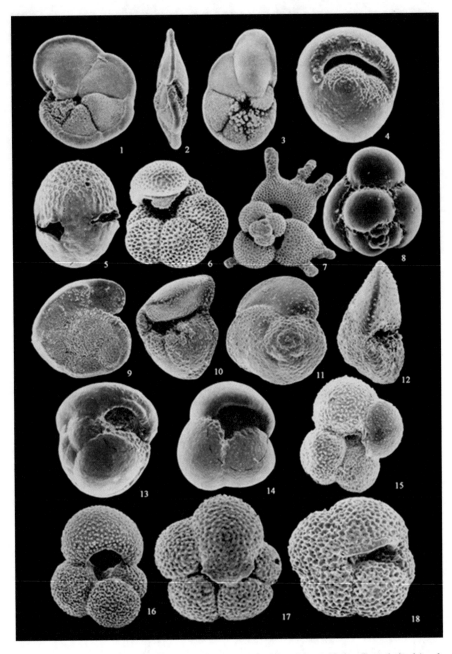

Fig. 6.13 Biogeographic distribution of recent planktic foraminifera: (Low-latitude): *1. Globorotalia menardii* X55; *2. Globorotalia menardii* X55; *3. Globorotalia tumida* X44; *4. Pulleniatina obliquiloculata* X89; *5. Sphaeroidinella dehiscens* X77; *6. Neogloboquadrina dutertrei* X113; *7. Globigerinoides fistulosus* X44; *8. Candeina nitida* X102. (Mid-latitude): *9. Globorotalia truncatulinoides* X80; *10. Globorotalia truncatulinoides* X99; *11. Globorotalia hirsuta* X73; *12. Globorotalia hirsuta* X99; *13. Globorotalia inflata* X110; *14. Globorotalia inflata* X88; *15. Globigerina falconensis* X159. (High-latitude): *16. Globigerina bulloides* X142; *17. Globigerina quinqueloba* X203; *18. Neogloboquadrina pachyderma* X183*

tubular and coiled forms. *Platysolenites* is believed to be the rootstock from which foraminifera diversified in the Phanerozoic. The Early Cambrian forms are morphologically fairly advanced and it has long been believed that simple ancestors of Cambrian foraminifera would have evolved in the Neoproterozoic, but would not have been adequately preserved. Molecular studies in the past decade have established the presence of living naked foraminifera and suggest a large radiation of non-fossilized naked and unilocular species which diverged from a Cercozoan ancestor in the Neoproterozoic, around 690–1150 Ma (Pawlowski et al. 2003).

The early geologic history of foraminifera has revealed a progressive transformation of the test from primitive organic-walled to agglutinated, and finally to secreted calcareous walled. The porcelaneous wall developed from the microgranular type that, in turn, had developed from the organic wall. Again, molecular studies have different interpretations, both the organic and agglutinated walls branching together at the base of the foraminiferal tree. It does, however, suggest that the porcelaneous type was derived from a more evolved agglutinated or calcareous lineage (Pawlowski et al. 2003).

After the initial diversification of unilocular, organic to agglutinated wall foraminifera in the Early Cambrian, the multilocular forms evolved in the Late Cambrian. A major biological change in foraminifera occurred in the late Paleozoic, when they attained a large size and complex internal morphology, constituting fusulinids that dominated the invertebrates of Carboniferous and Permian times. Several genera of the fusulinids were greater than 1 cm in size and some are reported to have reached a length of 14 cm. The living large-sized foraminifera occur in shallow water and host algal symbionts. The fusulinids would also have adapted to a symbiotic mode of life and, thus, photosymbiosis represents a major event in the evolutionary history of foraminifera. All the microgranular-walled foraminifera became extinct at the end of the Permian.

Foraminifera adapted to a planktic mode of life in the early Jurassic. It is postulated that dissociation of gas hydrate and the consequent oceanic anoxic event in the early Toarcian (Late Triassic) probably caused the transition of benthic foraminifera to a planktic mode of life. *Oberhauserella quadrilobata* and *Praegubkinella* are benthic or quasi-planktic foraminifera that evolved into the planktic genus *Conoglobigerina* in the Early Jurassic, Bajocian Stage (Hart et al. 2003). *Globuligerina oxfordiana* is the most widely recorded species of the Jurassic age, occurring in several localities in Europe, Russia and Canada. Although evolved in the Early Jurassic, planktic foraminifera diversified and gained importance in the Early Cretaceous. *Heterohelix*, *Praeglobotruncana*, *Rotalipora*, *Schackoina* and *Ticinella* appeared in Aptian–Albian times. Further diversification occurred in the Late Cretaceous, but the episode of mass extinction at the end of the Cretaceous eliminated them all. *Globotruncana*, *Hedbergella*, *Heterohelix*, *Praeglobotruncana*, *Rotalipora*, *Marginotruncana* and *Archaeoglobigerina* were the common planktic genera of this time (Fig. 6.14). Stable isotope data suggest that a number of Cretaceous planktic foraminifera were strongly photosymbiotic. These included the species of *Pseudoguembelina*, *Heterohelix*, *Racemiguembelina* and *Planoglobulina* (D'Hondt and Zachos 1998).

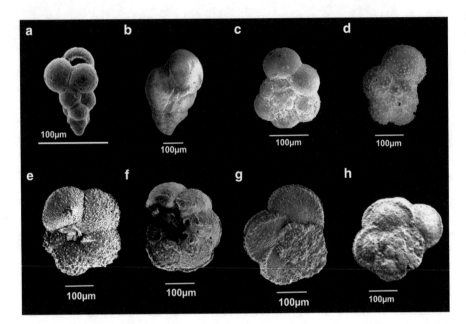

Fig. 6.14 Cretaceous planktic foraminifera (**a**) *Guembelitria* (**b**) *Heterohelix* (**c**) *Hedbergella* (**d**) *Whiteinella* (**e**) *Praeglobotruncana* (**f**) *Rotalipora* (**g**) *Marginotruncana* (**h**) *Globotruncana* (courtesy, C N Ravindran)

The trend of planktic foraminiferal extinction at the K/T boundary has been well studied in the El Kef section of Tunisia (Keller 1988). It is observed that extinction begins below the impact event (marked by an iridium anomaly) in the section, and there is a sequential elimination of less tolerant species. The complex and large species disappear first, followed by smaller and less ornate species, and the primitive, smallest species survive the longest. The evolution of species after the K/T event occurs in two pulses. Small, primitive and less diverse species appear first, followed by the large and more diverse species.

The Cenozoic is marked by the most striking evolutionary development in the history of both benthic and planktic foraminifera. The informally known larger benthic foraminifera (LBF) that had evolved repeatedly in the late Paleozoic and Cretaceous appeared again in great abundance and with widespread distribution in the Paleogene (Fig. 6.15). Although not as abundant and widespread, the modern LBF have provided a good insight into the biology and ecology of this group of foraminifera. Their characteristics include: (1) large size, (2) algal symbiosis, (3) occurrence in tropical to subtropical latitudes, (4) habitation of shallow waters of oligotrophic seas, and (5) longevity of a few months to 2 years. Delayed maturation and growth to large size are advantageous under limited food resources and stable environmental conditions. These traits, along with algal symbiosis, ensured the success of these species in warm, shallow, stable and oligotrophic environments (Hallock 1985). A close similarity is observed in the evolutionary trend of Paleogene larger benthic and planktic foraminifera. The specialist planktic morozovellids and acarinids and large-sized, internally complex benthic

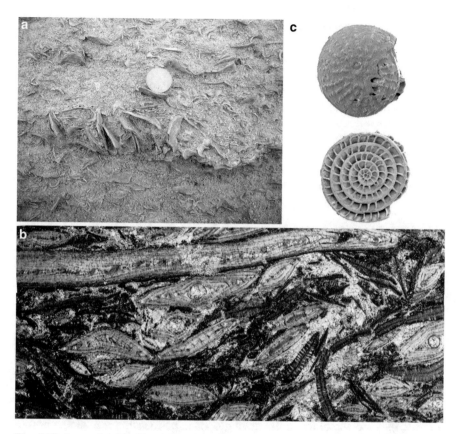

Fig. 6.15 Larger benthic foraminifer *Nummulites* formed thick sequence of limestone in the Eocene: a close-up view in an outcrop (**a**), associated with *Discocyclina* in a thin section under an optical microscope (**b**) and under a scanning electron microscope (**c**)

foraminifera both diversified in the Paleocene and early Eocene in response to an abundance of niches in oligotrophic surface waters of low to mid-latitudes (Hallock et al. 1991). The global warming at the close of the Paleocene and the onset of the Eocene was associated with turnover of larger foraminifera, characterized by an increase in test size, adult dimorphism and an increase in the specific diversity of these foraminifera. The enhanced oligotrophy from the Late Paleocene onward is believed to have favoured the rapid expansion of the k-strategist LBF (Scheibner et al. 2005).

The general morphologic characteristics of planktic foraminifera underwent marked changes across the K/T and Eocene/Oligocene extinctions. Morphologically complex tests made way for simple forms after the extinctions. Tests with keels, apertural lips, secondary calcification and rugose surfaces were replaced by simple trochospiral tests with smooth surfaces (Lipps 1986). The diversification of planktic foraminifera in the Cenozoic, after the K/T extinction, is tested by a growth model (Collins 1989) that suggests an explosive initial diversification rate for less than 1 Ma, followed by a slow, fluctuating growth to a Middle Eocene maximum of 44 species. The rest of the Paleogene was marked by a very slow decline in species richness. No noticeable correlation was found between temperature fluctuation and global diversity. The planktic foraminiferal families Globigerinidae and Globorotalidae appeared in the Paleocene and later diversified in the rest of the Cenozoic. Photosymbiosis recurred after the Cretaceous extinction and evolved in many species, including the living *Hastigerina pelagica* and *Orbulina universa*. A large-scale survey of the marine biodiversity in different oceanic environments has revealed worldwide consistency, despite obvious differences in environmental conditions of the various oceanographic regimes. In general, there is an increase in marine biodiversity during the Cenozoic and, more strikingly, with the advent of the Neogene.

6.2   Ostracoda

Ostracodes are microscopic crustaceans forming a distinct Class Ostracoda in the Phylum Crustacea. Carl Linne in 1758 and O.F. Muller in 1776 were the first to describe ostracodes, and since then, more than 65,000 living and fossil species have been reported. Ostracodes have bivalve carapaces that are weakly to strongly calcified. The carapaces are usually 0.5–2.0 mm long, but some forms reach a size of 30 mm. They are benthic, nektobenthic and pelagic. They are one of the most successful groups to colonize diverse environments from deep marine to shallow marine, lakes and fresh water to damp, terrestrial environments. Salinity and temperature are principal environmental controls in the distribution of ostracoda, and as a result, they are one of the most useful groups in paleoenvironmental reconstruction. The chemistry of the ostracoda carapace provides valuable information about hydrology and precipitation, particularly of lake sediments that lack other calcareous shells. It has a long geological history of more than 400 million years, since it evolved in the Ordovician.

Morphology

Ostracodes have the typical body plan of the crustaceans, comprising head, thorax and pairs of appendages. The body is enclosed in a cuticle and its fold on either side of the body secretes a bivalve shell, known as a carapace. Various appendages protrude between the valves for locomotion, feeding and reproduction. The carapaces are generally bean-shaped, but their shapes may be extremely varied, including inflated, compressed, spheroidal and rectangular. The ostracodes grow by moulting, and the younger carapaces (instars) are preserved in the sediments associated with the adult and fully grown carapaces. The moult stages (or instars) are designated either as first, second, third, etc. (in ascending order of size), or as A (adult), A-1, A-2, etc. (in descending order). One group of ostracodes (the podocopids) has nine instars, while the other (myodocopids) consists of four to seven juvenile instars and an adult instar. Most marine ostracodes reproduce sexually, but non-marine ostracodes more commonly reproduce asexually (parthenogenesis).

The orientation of extant groups of ostracodes is easier, but it is problematic and uncertain in extinct groups. There are several criteria for determining the orientation of valves. The hinge is located on the dorsal margin. The ventral margin is often concave and may have prominent spines, flanges and wing-like alae that generally point towards the posterior. The adductor muscle scars are situated in the anterior half of the valve. The greatest width of the carapace is posterior and the greatest height in living ostracodes is anterior in position. Sulcus in several forms is in the anterior position.

The wall of the carapace consists of an outer and inner lamella, and the contact between the two is the line of concrescence (Fig. 6.16). The edge of the inner lamella is the duplicature. The free margin of the duplicature forms a space inside the ventral margin, called the vestibule, to house digestive and reproductive organs. The scars of adductor and mandibular muscles characterize the inside of the valve. These are diagnostic features in the taxonomic classification. Two types of pore canals characterize the valves of ostracoda. The normal or lateral pore canals are transverse to the outer lamellae and they are generally simple pores or sieve-like pores. The radial or marginal pore canals originate at the line of concrescence and extend through the valve to its outer margin. Sensory hairs protrude through the pores. The pores may flush with the surface, and may be funnel-shaped or open at the tips of conical spines. The characteristics of pores are useful in taxonomy.

The contact margin of the valves is divided into the cardinal or hinge margin and the extra-cardinal or free margin. The line along which the valves articulate is the hinge line. The following categories of hinge are distinguished:

Merodont hinge—possessing three elements, anterior and posterior teeth (or sockets) separated by the median bar (or groove).

Amphidont hinge—four-element hinge, anterior region of the median bar (or groove) differentiated and all elements may be crenulated.

Adont hinge—lacks teeth and sockets but has a single element comprising a groove in one valve and a corresponding ridge on the other.

The external surface of the valves may be smooth or ornamented with pits, reticulation, spines, tubercles and nodes (Fig. 6.17). Although the pattern of ornament within a species usually remains unchanged, the degree of ornamentation may be related to environmental parameters, including salinity, nature of substrate and organic carbon content of the sediments. Further, sexual dimorphism causes variation in shape, size and ornamentation of carapaces. The male carapaces may be larger or smaller than those of their female counterparts and the carapaces of females of some species are bulged in the antero-ventral part (due to an internal brood pouch). Kesling (1951) has described the terminology of carapaces and compared the terms used by different workers.

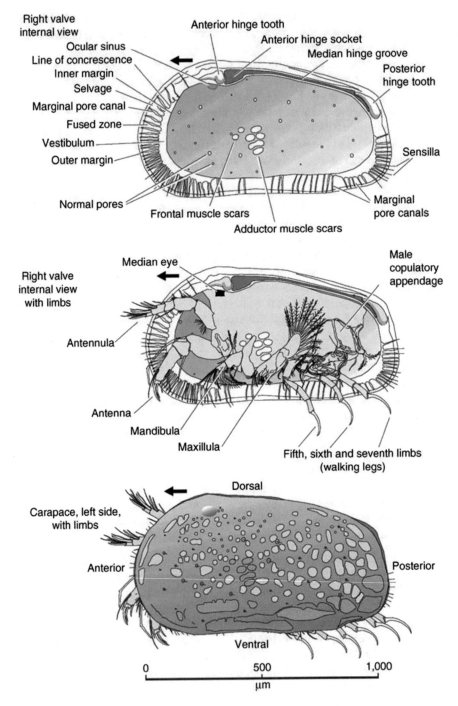

Fig. 6.16 Morphology of ostracode carapace (after Cronin 2009, in turn Courtesy D J Horne; reproduced with permission © Springer Science + Business Media)

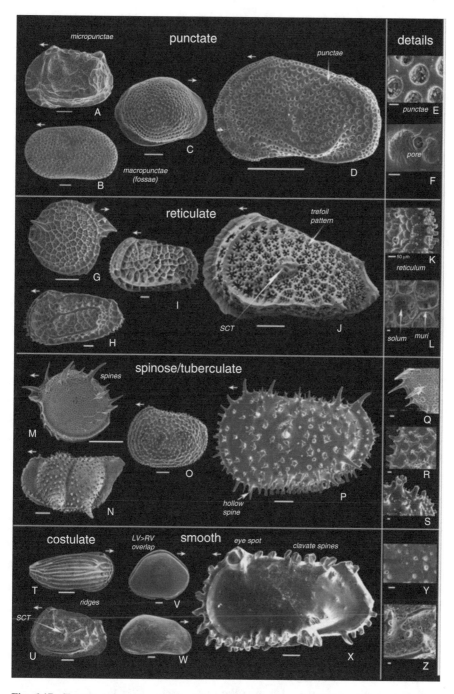

Fig. 6.17 Characteristic ornamentation of ostracod—punctuate, reticulate, spinose/tuberculate and smooth carapaces (from *top* to *bottom*) (after Rodriguez-Lazaro and Ruiz-Munoz 2012, reproduced with permission © Elsevier)

Classification

There are several schemes of classification of ostracoda. Biologists use soft part morphology, mainly the appendages, to classify the modern ostracodes, but these are not useful for the fossil ostracodes. The carapace morphology useful in classification includes the basic shape of the carapace, surface ornamentation, type of hinge, muscle scar morphology and pore characteristics. There are vigorous debates over phylogeny and the higher classification of ostracodes. The classification of Quaternary and Recent ostracoda is presented in Fig. 6.18.

Ecology

The living ostracodes occupy a range of aquatic to semi-aquatic environments. They occur at all depths, from the shallowest part of the ocean to the abyssal plain, in brackish and fresh waters, with some even occurring in the mosses and leaf litters of terrestrial environments. The myodocopids are almost exclusively marine, and range from benthic to nektobenthic and pelagic forms. These are predators, scavengers, detritus feeders and filter feeders. The podocopids are the most diverse and widespread ostracodes, living in virtually all types of marine, brackish and freshwater environments. The superfamily Cytheroidea is essentially marine, while Darwinulinoidea and Cypridoidea are freshwater. The majority of living brackish water ostracodes are podocopids and assigned to the following groups (Smith and Horne 2002):

1. Essentially marine taxa that are able to tolerate reduced and/or fluctuating salinities: *Semicytherura, Leptocythere, Propontocypris.*
2. Essentially non-marine taxa that can tolerate increased salinities: species of *Darwinulina* and *Candona.*
3. Truly brackish water taxa that may tolerate but are not usually found in fresh waters: species of *Loxoconcha, Leptocythere.* The brackish water species *Cyprideis torosa* is known to survive in abnormally high salinity of >50‰.

There are ecophenotypic variations in the shell morphology of ostracodes of marginal marine environments, mainly in response to salinity change. As reviewed by Anadon et al. (2002), the variations are reflected in the size of the carapace, ornamentation and shape of pores:

- Variation in salinity results in change of size such that a decrease in salinity leads to a size reduction in marine euryhaline species and an increase in size in freshwater euryhaline ones. Such relationships are recorded in *Leptocythere castanea* and *Loxoconcha impressa.*
- Some species, such as *Cytheromorpha paracastanea* (Recent), *Cytheromorpha ouachitaensis* (Eocene) and *Loxoconcha nystiana* (Oligocene), show weakening of ornamentation with a decrease in salinity. The opposite trend is observed in the freshwater euryhaline species *Limnocythere floridensis.*

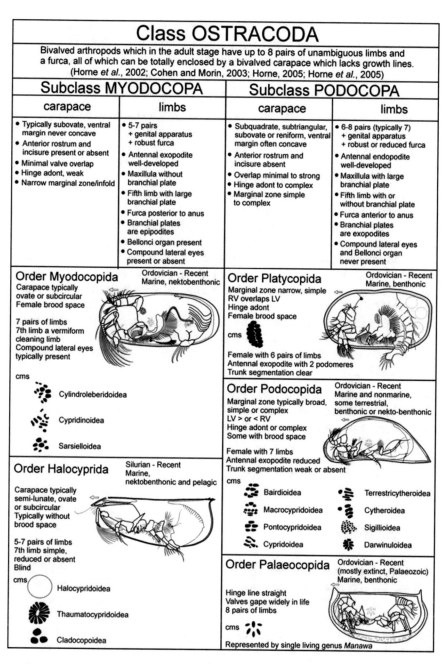

Fig. 6.18 Classification and characteristics of Quaternary and living ostracodes (after Rodriguez-Lazaroand and Ruiz-Munoz 2012, reproduced with permission © Elsevier)

- The thickness of valves in *Cyprideis torosa* varies with salinity, the thicker valves appearing in oligohaline waters and the thinner ones in hyperhaline waters.
- In *Cyprideis torosa*, the percentage of different shapes of sieve pores (round, elongate or irregular) on the same shell are linked to the salinity of the water and have been applied to infer paleosalinity variation in geological records.

Some species of ostracode tolerate a wide range of temperature while others are stenothermic, with narrow tolerance for temperature change. The cool water, deep-sea species are psychrospheric and warm water, shallow marine species are thermophilic. As discussed in the following section, the sensitivity of the ostracodes to the combined effects of salinity and temperature influenced the evolution of the group in the Cenozoic era. In the marine environment, the nature of the bottom sediments also has a considerable effect on the distribution of ostracodes. Coarsely ornamented carapaces are characteristic of coarse bottom sediments, while smooth or subdued ornamentations on carapaces occur on fine grained bottom sediments. The thick, highly ornamented *Carinocythereis*, *Costa* and *Mutilus* inhabit coarse grained sediments of near-shore environments, and thin, smooth-shelled *Cytheropteron*, *Erythrocypris*, *Krithe*, *Paracytherois* and *Xestoleberis* are found in offshore environments (Puri 1971).

A good relationship between tide levels and ostracode species is found in several places that suggests potential use of ostracodes in reconstruction of former sea levels. The main ostracode species associated with different tide levels are as follows (Boomer 1998):

Highest astronomical tide: No ostracoda.
Mean high water: *Loxoconcha elliptica*, *Leptocythere porcellanea*, *L. castanea*.
Mean high water neaps: *Leptocythere porcellanea*, *L. casertosa*, *L. baltica*, *Xestoleberis* sp.
Mean sea level (down to mean low water neaps): *Leptocythere pellucida*, *Loxoconcha rhomboidea*, *Hemicythere villosa*.

Geologic History

The extant subclasses of ostracoda, the Podocopa and Myodocopa, evolved in the Ordovician. There are Cambrian fossils attributed to ostracoda, the Phosphatocopida, but these are now excluded from the ostracoda. It is suggested that the expansion of ecological niches by Early Ordovician marine transgression favoured the first major radiation of ostracoda. The earliest known ostracoda were marine. The most abundant during the Paleozoic were Paleocopida, but they became virtually extinct by the end of the Permian. Leperditicopida is the only extinct order that disappeared at the end of the Devonian. The ostracodes invaded freshwater for the first time in the Early Carboniferous, and Darwinulocopina flourished during the Carboniferous, Permian and Triassic. The family Darwinulidae have reproduced exclusively by parthenogenesis (asexually) for over 200 million years. By the early Triassic, the majority of palaeocopids had become extinct. The ostracodes exploited deltaic, marginal lagoon and lacustrine environments by the Late Jurassic and Early Cretaceous. Stratigraphically important genera that originated in the Early Cretaceous are *Neocythere*, *Cythereis* and *Platycythereis* and those that originated in the Late Cretaceous include *Krithe*, *Brachycythere*, *Mauritisina* and *Trachyleberidea*. The Trachylebiridids and their derivative families became important elements of the Cenozoic assemblage.

The psychrospheric ostracode fauna developed in the Late Eocene (40–38 Ma) with the formation of cold Antarctic bottom water, closing of the Tethyan seaway and establishment of thermohaline circulation of the world oceans. The closing and opening of oceanic gateways have played an important role in ocean circulation patterns and climate, both of which influenced the general distribution of the ostracode fauna. The deep-sea ostracoda have indicated the following global paleoceanographic events in the Cenozoic (Benson 1990):

1. Late Eocene (40–38 Ma): Development of psychrospheric ostracode fauna, ornate, reticulate genera comprising *Bradleya*, *Poseidonamicus* and *Agrenocythere*.
2. Middle Miocene (16–14 Ma): Increase in population and diversity of ostracodes in the Indo-Pacific and South Atlantic, concomitant with major Antarctic ice sheet formation.
3. End of the Miocene (6.3–4.9 Ma): Messinian salinity crisis, occurrence of psychrospheric ostracode fauna at shallow depths. The Pliocene boundary event is characterized by an invasion of *Agrenocythere pliocenica* and *Oblitacythereis mediterranea* in the Mediterranean, the species that occurred bathyal Atlantic in the Miocene.
4. Pliocene (3.5 Ma): Sharp decline in generic diversity of ostracode fauna, as intense as at the end of the Cretaceous when 15 % of the genera became extinct.

The ostracode genera that originated in the Cenozoic and continue to the present day include *Pokornyella*, *Cytherella*, *Bradleya*, *Abyssocythere*, *Henryhowella*, *Agrenocythere*, *Hemicythere* and *Cythere*, among others.

6.3 Calcareous Nannoplankton: Coccolithophores

The calcareous nannoplankton are plankton <63 μ in size that have calcareous tests. The Coccolithophores and Discoasters are examples of calcareous nannoplankton of geological importance. The coccolithophores (or coccolithophorids) are autotrophic nannoplankton belonging to the Haptophyta division (unicellular alga) and are generally of <20 μ size. The test of the coccolithophores is known as a coccosphere and the tiny calcareous plates constituting it are called coccoliths. *Emiliania huxleyi* is a well-known and extensively studied living coccolithophore, widely distributed in pelagic and near-shore areas of tropical to higher latitudes. By their sheer abundance, the coccolithophores and other nannoplankton form calcareous oozes in present-day deep sea, and constituted rocks, called chalk, in the geologic past. The nannoplankton oozes occur at slightly greater depths (near the CCD) compared with foraminiferal oozes because of nannoplankton being more resistant to dissolution

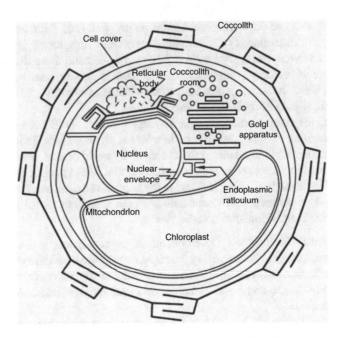

Fig. 6.19 Schematic representation of the cell structures of coccolithophores

than the planktic foraminifera. The coccolithophores appeared in the Late Triassic, diversified in the Early Jurassic and became a biostratigraphically significant group of microfossil in the Cenozoic. The discoasters are larger in size and their easily observable morphological features make them a good microfossil for age determination. Due to the minuteness of the coccoliths, the identification of taxa requires examination under a high power optical microscope or, even better, a scanning electron microscope. The study of coccolithophores was popularized mainly due to application of the electron microscope and the recovery of deep-sea cores by the DSDP. Due to their small size, good recovery, even from sidewall cores, and short stratigraphic ranges, the calcareous nannofossils have become a reliable tool for the biostratigraphic dating of both Mesozoic and Cenozoic sediments.

Biology of the Coccolithophore

The cell of the coccolithophore is of a spherical or oval shape, having a nucleus, golgi apparatus and two golden-brown chromatophores (Fig. 6.19). Whip-like organs, flagella and haptonema facilitate the movement of the cell. The chloroplasts contain the chlorophyll for photosynthesis. The calcitic coccoliths are formed in cellular vesicles and are transported outside to the cell surface. A reticular body consisting of sub-parallel strands, the golgi body and nucleus are also connected with the formation of coccoliths. The functions of coccoliths include (1) protection of cell against damage and predation, (2) regulation of light allowed into the cell, and (3) flotation by adopting a suitable shape (elongate coccosphere, for example, contribute to buoyancy). In a living coccolithophore, *Coccolithus pelagicus*, two phases of life cycle are recognized. The motile phase possessing flagellar apparatus alternates with the non-motile phase. The motile phase is either naked or bears coccoliths of a different type than the non-motile phase that secretes coccoliths. Early on, many viewed the two phases as different species. The coccolithophores photosynthesize and, therefore, they are at the bottom of the marine food chain and a source of food for the herbivorous bacteria. Reproduction in coccolithophore is either by fission of the mother cell and subsequent regeneration of coccoliths or by repeated divisions of the mother cell inside the crust of coccoliths and release of motile or non-motile daughter cells. These may or may not regenerate coccoliths.

Morphology

There are two types of coccoliths, the holococcoliths and the heterococcoliths. The holococcoliths are made of numerous minute crystallites, all rhombohedral or hexagonal crystals of similar shape and size. The crystallites in heterococcoliths, contrastingly, are of variable shape and size. Structurally, the heterococcoliths are more rigid than the holoccoliths. The holoccoliths, therefore, disintegrate easily during post-mortem sinking to the seafloor. The heterococcoliths are built of plates, rods and grains. The elliptical or circular shields are constructed of radial plates, the central area of which may be empty or have crossed bars and produced into a spine (Fig. 6.20). *Coccolithus pelagicus* secretes holococcoliths in the motile phase and heterococcoliths in the non-motile phase of its life cycle. Mineralogically, the coccoliths are normally calcite and sometimes aragonite or vaterite. Due to their unstable nature, aragonite and vaterite are not found in fossil coccoliths. The classification of coccolithophorids is based on coccolith morphology, the mode of arrangement of coccoliths on the cell and the shape of the cell.

The nannofossil Discoaters are of star or rosette shape. They became extinct near the Pliocene—Quaternary boundary and, therefore, the exact mechanism of their calcification is unknown. Many believe that discoasters remained within the cell after calcification and secreted around it. They secreted tabular calcite, unlike the rhombohedral or hexagonal crystals of the coccoliths, and were more coarsely constructed. Because of their robust construction, discoasters are more resistant to dissolution. Young et al. (1997) have presented standardized terminology with illustrations for the coccoliths and other calcareous nannofossils.

Ecology

The coccolithophorids mostly live in marine waters, with some occurring in fresh and brackish waters. *Emiliania huxleyi* tolerates a salinity range of 45–15‰. *Hymenomonas roseola* is the only freshwater species known today. Due to the

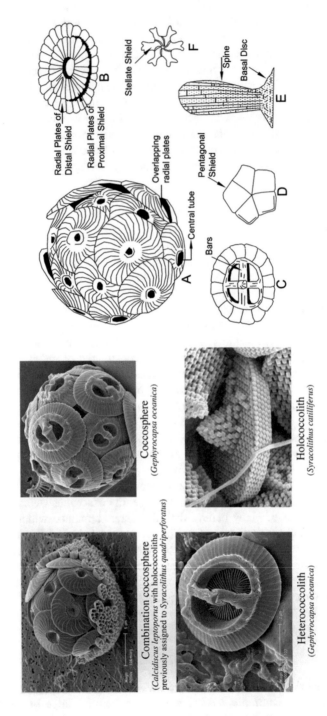

Fig. 6.20 Morphology of coccolithophores illustrating the general terminology used to describe them. (SEM micrographs of coccoliths after Baumann et al. 2005, reproduced with permission © Springer Science+Business Media)

requirement of photosynthesis, their vertical distribution in the ocean is limited to the photic zone, which is the uppermost 200 m of the water column. Both the modern and the fossil calcareous nannoplankton suggest their high diversity in low latitudes, although they also occur in high latitudes. Discoasters were mostly adapted to tropical and warm subtropical waters. Their abundance is an indicator of warm surface waters. Temperature is an important factor in the growth of coccoliths. Optimum growth of *E. huxleyi* in laboratory culture is observed within 18–24 °C. This species has, however, the widest biogeographic and temperature range among all the coccolithophores. The following biogeographic distribution in the Pacific suggests the temperature preferences of different species (Fig. 6.21; McIntyre et al. 1970):

Tropical (above 29 °C):	*Umbellosphaera irregularis, Gephyrocapsa oceanica* and *Emiliania huxleyi* constitute nearly 100 % of the total flora
Tropical (>21 °C):	*Umbellosphaera irregularis, Gephyrocapsa oceanica*
Subtropical (14–21 °C)	*Gephyrocapsa ericsonii, Rhabdosphaera stylifera, Discosphaera tubifera*
Transition (6–14 °C):	*Gephyrocapsa caribbeanica, Coccolithus pelagicus, Emiliania huxleyi*
Subarctic (6–14 °C):	*Emiliania huxleyi*

Among nutrients, nitrate is essential for the growth and calcification of coccolithophores, but unnaturally high nutrient inhibits calcification. The highest abundance of coccolithophores within phytoplankton communities is in oligotrophic environments (Baumann et al. 2005). Young (1994) recognized three ecological communities of coccolithophores as follows:

1. Placolith-bearing species (coccoliths composed of proximal and distal shields joined by a central column): characterize coastal or mid-ocean upwelling regions; example, *Emiliania huxleyi, Gephyrocapsa, Umbilicosphaera*.
2. Umbelliform assemblages (coccoliths with large processes which flare distally to produce a double-layered coccosphere): characterize oligotrophic, mid-ocean environments; example, *Umbellosphaera, Discosphaera tubifera*.
3. Floriform species (dense asymmetrical mass of coccoliths): characterize the deep photic zone in stable water columns of low to mid-latitudes; example, *Florisphaera profunda*.

Geologic History

The calcareous nannofossils originated in the late Triassic and became an important part of the marine plankton community with two major radiations, first in the Early Jurassic and second in the early Paleogene. There were two major events in the Mesozoic that caused calcareous nannofossils to become important rock-building components. The diversity and abundance increased markedly in the Tithonian–Berriasian

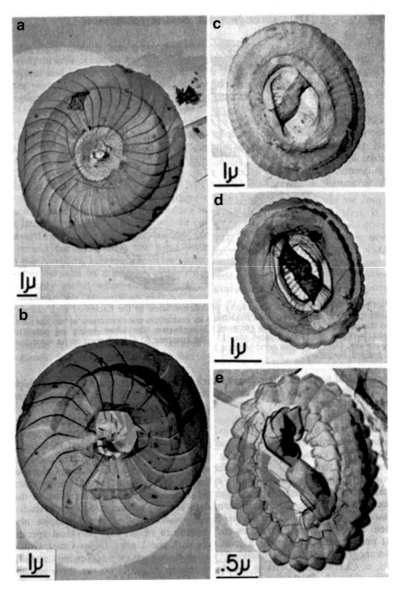

Fig. 6.21 Latitudinal distribution of selected coccolithophores in the Pacific. (**a**) *Cyclococcolithus leptoporus* var B (tropical, warm water), (**b**) *Cyclococcolithus leptoporus* var C (eurythermal, extending to warm subpolar), (**c**) *Gephyrocapsa oceanica* (tropical), (**d**) *Gephyrocapsa caribbeanica* (most cosmopolitan, tropical to subpolar), (**e**) *Gephyrocapsa ericsonii* (subtropical) (after McIntyre et al. 1970, reproduced with permission ©John Wiley and Sons)

and in the Late Cretaceous when, for the first time in the history of the earth, carbonate production was substantially influenced by planktic organisms. It resulted in the widespread distribution of chalky sediments in epicontinental seas all over the globe. A synthesis of the biogeographic distribution suggests the following assemblages in the Cretaceous (Mutterlose et al. 2005):

High latitudinal assemblage: *Watznaueria barnesae, Crucibiscutum salebrosum, Sollasites horticus.*

Mid–low latitudinal assemblage: *Watznaueria barnesae, Biscutum constans, Zeugrhabdotus.*

Low latitudinal assemblage: *Watznaueria, Rhagodiscus asper, Micrantholithus, Conusphaera.*

A study of evolutionary indices of calcareous nannoplankton across the Meso-Cenozoic (Bown 2005) provides important information about their evolution. The study concludes that the species richness, evolutionary rates and species longevity of coccolithophore all show a marked difference above and below the K/T boundary. In comparison to the Mesozoic, the species richness was highly variable in the Cenozoic and the species decline was significant in the Oligocene and Pliocene–Pleistocene. The marked difference in the evolutionary indices of coccolithophores between the Mesozoic and the Cenozoic is attributed to a shift in climate mode from greenhouse in the Cretaceous and early Paleogene to icehouse in the Oligocene. Most of the Mesozoic biodiversity was established in the Early Jurassic during relatively warm climate, high sea level and flooded continental shelves.

6.4   Pteropods

Pteropods are marine, pelagic gastropods having aragonite shells. Their size usually ranges between 0.3 and 10 mm. The aragonitic shells are more susceptible to solution than the calcitic tests of foraminifera or calcareous nannoplankton and, therefore, have poor preservation potential. The thin and fragile shells of pteropods further limit their preservation in pre-Quaternary geological records. In view of this, the stratigraphic utility of this group is limited. The pteropods are known with certainty to have existed from the Cretaceous onwards. The influence of temperature, salinity and productivity of their distribution, however, makes them useful in paleoceanography and the paleoenvironmental reconstruction of Quaternary successions. Pteropods contribute as high as a quarter of the calcium carbonate on the seafloor, with a maximum flux at the end of the southwest monsoon season (Singh and Conan 2008). The high abundance of pteropods contributes to pteropod oozes in the deep sea, above the aragonite compensation depths. These occur in the Atlantic at depths of less than 3000 m and in even shallower levels in the Pacific and Indian Oceans.

All pteropods do not have shells and, therefore, only a portion of the living pteropods have left fossil records. The shells are generally trochospiral, planispiral, discoid, cylindrical and elongate cones (Fig. 6.22). There are only a few families that

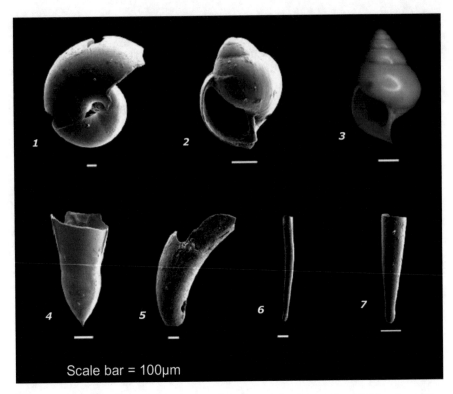

Scale bar = 100μm

Fig. 6.22 Quaternary pteropods from the Arabian Sea: (*1*) *Limacina inflata* (*2*) *Limacina trocho-formis* (*3*) *Limacina bulimoides* (*4*) *Clio convexa* (*5*) *Cavolinia gibbosa* (*6*) *Creseis acicula* (*7*) *Creseis virgula* (reproduced after Rai et al. 2008, with permission from Indian Academy of Sciences and the author)

have shell-bearing representatives. The family Peraclididae is characterized by a sinistrally coiled shell, twisted prolongation of the columella, and delicate reticulation on the surface. The family Cavolinidae comprises uncoiled shells and is without operculum. The coiled genera *Limacina* and *Peraclis* are thought to be primitive, while the uncoiled genera belonging to Cavolinidae are advanced forms.

Pteropods are exclusively marine organisms inhabiting the upper 500 m of the water column. Temperature has a major influence on their distribution and many species have a limited tolerance for it. The cold, polar regions are inhabited only by the species *Limacina helicina*. The cold-temperate species include *Limacina retroversa* and *Clio pyramidata pyramidata*. Several species occur in the warm water of the tropical regions, comprising the species *Cavolina, Creseis, Clio, Limacina, Styliola* and *Diacria*. The water-depth-controlled life strategy of many pteropod species makes them useful paleobathymetric indicators, particularly over the continental margins (Singh et al. 1998). The epipelagic *Creseis* is most abundant in recent sediments <50 m in depth, the mesopelagic *Limacina inflata* between depths of 50 and 100 m, and the *Clio convexa* at water depths >100 m on the western continental shelf of India. The distribution patterns of pteropods in this region are controlled by a combination of factors, including bathymetry, hydrography and conditions of aragonite preservation (Singh et al. 2005).

Primary productivity is another important ecological factor in the distribution of pteropods. An understanding of the present-day distribution of pteropods in relation to productivity has led to the reconstruction of productivity changes in glacial-interglacial intervals. *Limacina inflata* is abundant in the present-day oligotrophic Gulf of Aqaba. It was also a dominant taxon in the last interglacial. *Limacina trochiformis* characterizes highly productive areas, and its abundance in the glacial intervals of Gulf of Aqaba cores indicates the increased fertility condition of the water (Almogi-Labin 1982). It is also noted that pteropod preservation is influenced by fluctuation in the oxygen minimum zone due to deep-sea ventilation and remains largely independent of variation in surface productivity (Rai et al. 2008). The low abundance of pteropods may, therefore, be due to dissolution of their solution-susceptible aragonite shell and not low productivity.

In recent years, an exciting application of pteropods has been found in understanding the effects of rising temperature and partial pressure of CO_2 (pCO_2) and the consequent ocean acidification on calcifying marine organisms. Due to their solution-susceptible aragonite shells, the pteropods are most vulnerable and likely to be the first major group adversely affected by ocean acidification. In an experimental study on a polar species *Limacina helicina*, it is observed that (1) temperature and pCO_2 had a significant effect on the mortality of the species, with temperature being the overriding factor, (2) the growth of the shell was significantly impacted and shell degradation, including corrosion and perforations, increased significantly with the rising pCO_2, and (3) the rising temperature and pCO_2 may result in a possible decline in abundance of the pteropods (Lischka et al. 2011).

References

Almogi-Labin A (1982) Stratigraphic and paleoceanographic significance of late Quaternary pteropods from deep-sea cores in the Gulf of Aqaba (Elat) and northernmost Red Sea. Mar Micropaleontol 7:53–72

Anadon P, Gliozzi E, Mazzini I (2002) Paleoenvironmental reconstruction of marginal marine environments from combined paleoecological and geochemical analyses on ostracods. In: Holmes JA, Chivas AR (eds) The Ostracoda: applications in quaternary research, vol 131, Geophysical monograph. American Geophysical Union, Washington, DC, pp 227–247

Baumann K, Andruleit H, Böckel B, Geisen M, Kinkel H (2005) The significance of extant Coccolithophores as indicators of ocean water masses, surface water temperatures and palaeo-productivity: a review. Paläontol Z 79:93–112

Benson RH (1990) Ostracoda and the discovery of global Cainozoic palaeoceanographical events. In: Whatley R, Maybury C (eds) Ostracoda and global events. Chapman and Hall, London, pp 41–58

Bijma J, Farber WW, Hemleben C (1990) Temperature and salinity limits for growth and survival of some planktonic foraminifers in laboratory cultures. J Foraminifer Res 20:95–116

Boomer I (1998) The relationship between meiofauna (ostracoda, foraminifera) and tidal levels in modern intertidal environment of North Norfolk: a tool for palaeoenvironment reconstruction. Bull Geol Soc Norfolk 46:17–29

Bown PR (2005) Calcareous nannoplankton evolution: a tale of two oceans. Micropaleontology 51:299–308

Collins LS (1989) Evolutionary rates of a rapid radiation: the Paleogene planktic foraminifera. Palaios 4(3):251–263

Cronin TM (2009) Ostracodes. In: Gornitz V (ed) Encyclopedia of paleoclimatology and ancient environments. Springer, The Netherlands, pp 663–665

Culver SJ (1991) Early Cambrian foraminifera from West Africa. Science 254:689–691

D'Hondt S, Zachos JC (1998) Cretaceous foraminifera and the evolutionary history of planktic photosymbiosis. Paleobiology 24(4):512–523

d'Orbigny A (1826) Tableau méthodique de la classe des céphalopodes. Ann Sci Nat 7:96–245

Goldstein ST (1999) Foraminifera: a biological overview. In: Sen Gupta BK (ed) Modern foraminifera. Kluwer Academic, The Netherlands, pp 37–55

Hallock P (1985) Why are larger foraminifera large? Paleobiology 11(2):195–208

Hallock P, Premoli Silva I, Boersma A (1991) Similarities between planktonic and larger foraminiferal evolutionary trends through Paleogene paleoceanographic changes. Palaeogeogr Palaeoclimatol Palaeoecol 83:49–64

Hart MB, Hylton MD, Oxford MJ et al (2003) The search for the origin of the planktic foraminifera. J Geol Soc Lond 160:341–343

Hayward BH, Holzmann M, Grenfell HR et al (2004) Morphological distinction of molecular types in *Ammonia* – towards a taxonomic revision of the world's most commonly misidentified foraminifera. Mar Micropaleontol 50:237–271

Jorissen FJ, de Stigter HC, Widmark JGV (1995) A conceptual model explaining benthic foraminiferal microhabitats. Mar Micropaleontol 26:3–15

Kaminski MA (2004) The Year 2000 classification of the agglutinated foraminifera. In: Bubik M, Kaminski MA (eds) Proceedings of the sixth international workshop on agglutinated foraminifera, Vol 8 Grzybowski Foundation Special Publication. pp 237–255

Keller G (1988) Extinction, survivorship and evolution of planktic foraminifera across the Cretaceous/Tertiary boundary at El Kef, Tunisia. Mar Micropaleontol 13:239–263

Kesling RV (1951) Terminology of ostracod carapaces. Contributions from Museum of Paleontology, University of Michigan, IX(4): 93–171

Kroon D, Nederbragt AJ (1990) Ecology and paleoecology of triserial planktic foraminifera. Mar Micropaleontol 16:25–38

Lipps JH (1986) Extinction dynamics in pelagic ecosystems. In: Elliott DK (ed) Dynamics of extinction. John Wiley, New York, pp 87–104

Lischka S, Büdenbender J, Boxhammer T, Riebesell U (2011) Impact of ocean acidification and elevated temperatures on early juveniles of the polar shelled pteropod Limacina helicina: mortality, shell degradation and shell growth. Biogeosciences 8:919–932

Loeblich AR, Tappan H (1964) Protista 2 Sarcodina chiefly "Thecamoebians" and Foraminiferida, (2 Vols). In: Moore RC (ed) Treatise on invertebrate paleontology. Geological Society of America and University of Kansas Press, Lawrence, KS

Loeblich AR, Tappan H (1987) Foraminiferal genera and their classification (2 Vols). Van Nostrand Reinhold, New York

Loeblich AR, Tappan H (1992) Present status of foraminiferal classification. In: Takayanagi Y, Saito T (eds) Studies in benthic foraminifera. Tokai University Press, Tokyo, pp 93–102

McIlroy D, Green OR, Brasier MD (2001) Palaeobiology and evolution of the earliest agglutinated foraminifera: *Platysolenites*, *Spirosolenites* and related forms. Lethaia 34:13–29

McIntyre A, Be AWH, Roche MB (1970) Modern Pacific Coccolithophorida: a paleontological thermometer. Trans N Y Acad Sci 32(6):720–731

Murray JW (2007) Biodiversity of living benthic foraminifera: how many species are there? Mar Micropaleontol 64:163–176

Mutterlose J, Bornemann A, Herrle JO (2005) Mesozoic calcareous nannofossils – state of the art. Paläontol Z 79:113–133

Pawlowski J, Holzmann M (2002) Molecular phylogeny of Foraminifera – a review. Eur J Protistol 38:1–10

Pawlowski J, Holzmann M, Berney C et al (2003) The evolution of early foraminifera. Proc Natl Acad Sci U S A 100(20):11494–11498

Pawlowski J, Holzmann M, Tyszka J (2013) New supraordinal classification of foraminifera: molecules meet morphology. Mar Micropaleontol 100:1–10

Puri HS (1971) Distribution of ostracodes in the oceans. In: Funnell BM, Riedel WR (eds) The micropaleontology of oceans. Cambridge University Press, Cambridge, pp 163–169

Rai AK, Singh VB, Maurya AS, Shukla S (2008) Ventilation of northwestern Arabian Sea oxygen minimum zone during last 175 kyrs: a pteropod record at ODP site 728A. Curr Sci 94:480–485

Rodriguez-Lazaro J, Ruiz-Munoz F (2012) A general introduction to ostracods: morphology, distribution, fossil record and applications. In: Horne DJ, Holmes JA, Rodriguez-Lazaro J, Viehberg FA (eds) Ostracoda as proxies for quaternary climate change, vol 17, Developments in quaternary science. Elsevier, Amsterdam, pp 1–14

Röttger R (1992) Biology of larger foraminifera: present status of the hypothesis of trimorphism and ontogeny of the gamont of *Heterostegina depressa*. In: Takayanagi Y, Saito T (eds) Studies in benthic foraminifera, Proceedings of the fourth international symposium on benthic foraminifera, Sendai, 1990. Tokai University Press, Tokyo, pp 43–54

Scheibner C, Speijer RP, Marzouk AM (2005) Turnover of larger foraminifera during the Paleocene – Eocene thermal maximum and paleoclimatic control on the evolution of platform ecosystem. Geology 33(6):493–496

Sen Gupta BK (1999) Systematics of modern Foraminifera. In: Sen Gupta BK (ed) Modern foraminifera. Kluwer Academic, The Netherlands, pp 7–36

Singh AD, Conan SMH (2008) Aragonite pteropod flux to the Somali Basin, NW Arabian Sea. Deep Sea Res I 55:661–669

Singh AD, Rajarama KN, Ramachandran KK, Suchindan GK, Samsuddin M (1998) Pteropods as bathometers: a case study from the continental shelf off Kerala coast, India. Curr Sci 75:620–623

Singh AD, Nisha NR, Joydas TV (2005) Distribution patterns of recent pteropods in surface sediments of the western continental shelf of India. J Micropalaeontol 24:39–54

Smith AJ, Horne DJ (2002) Ecology of marine, marginal marine and non-marine ostracodes. In: Holmes JA, Chivas AR (eds) The Ostracoda: applications in quaternary research, vol 131, Geophysical monograph. American Geophysical Union, Washington, DC, pp 37–64

Streeter SS (1973) Bottom waters and benthonic foraminifera in the North Atlantic: glacial–interglacial contrasts. Quat Res 3:131–141

Young JR (1994) Function of coccoliths. In: Winter A, Seisser WG (eds) Coccolithophores. Cambridge University Press, Cambridge, pp 63–82

Young JR, Bergen JA, Bown PR et al (1997) Guidelines for coccolith and calcareous nannofossil terminology. Palaeontology 40:875–912

Further Reading

Armstrong HA, Brasier MD (2005) Microfossils, IIth edn. Blackwell, Oxford, UK

Cushman JA (1948) Foraminifera: their classification and economic use, 4th edn. Harvard University Press, Cambridge

Haynes JR (1981) Foraminifera. John Wiley, New York

Jones RW (2014) Foraminifera and their applications. Cambridge University Press, Cambridge

Lipps JH (1993) Fossil prokaryotes and protists. Blackwell, Boston

Sen Gupta BK (ed) (1999) Modern foraminifera. Kluwer Academic, The Netherlands

Horne DJ, Holmes JA, Rodriguez-Lazaro J, Viehberg FA (eds) (2012) Ostracoda as proxies for quaternary climate change, vol 17, Developments in quaternary science. Elsevier, Amsterdam

Chapter 7
Siliceous-Walled Microfossils

7.1 Radiolaria

Radiolarians are single-celled, marine zooplankton that can be found floating near the surface or at water depths of hundreds of metres. They belong to the Phylum Radiozoa. The size of radiolarian varies from 30 μm to 2 mm in diameter and some of them form macroscopic colonies that may reach a size of a centimetre or larger. Temperature, salinity and nutrient characteristics of the water masses control the distribution of radiolarian assemblages. The radiolarians, thus, preserve the signatures of oceanic and climatic changes of the past. Unlike calcareous microfossils, the radiolarians are preserved in deep-sea sediments independent of the CCD and have greater potentiality to be used as tools in biostratigraphy and paleoceanography, especially where calcareous microfossils are absent due to their dissolution (Sharma and Daneshian 1998). The radiolarians range from the Cambrian period, but the skeleton-less forms may have evolved in the Precambrian. They dominate the sediments below the carbonate compensation depth and form radiolarian oozes. Such oozes are mostly found in the equatorial Pacific below the zones of high productivity at depths of 3–4 km, but abundant radiolarians also occur at shallower depths associated with coccolith or planktic foraminiferal oozes. Fossil radiolarians have formed bedded cherts (radiolarites) in geologic records and radiolarite–ophiolite associations are known from subduction complexes.

Morphology

The living matter of the radiolarian cell comprises two layers: (1) a mucoid or chitinous sac in the centre, called the central capsule, holding the intracapsular cytoplasm and nucleus, and (2) the outer layer of cytoplasm surrounding the central capsule,

© Springer International Publishing Switzerland 2016
P.K. Saraswati, M.S. Srinivasan, *Micropaleontology*,
DOI 10.1007/978-3-319-14574-7_7

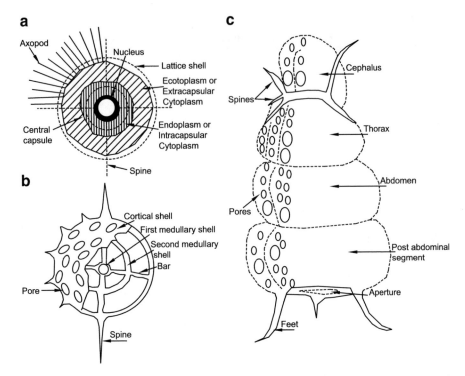

Fig. 7.1 General morphology and terminology of radiolarian cells (**a**) and spumellarian (**b**) and nassellarian (**c**) skeletons

known as the extracapsulum. Thread-like axopodia radiate from the central capsule and through the extracapsulum (Fig. 7.1). The intracapsular cytoplasm consists of the nucleus, vacuoles of varying sizes and the vital organelles such as mitochondria and Golgi bodies. The extracapsulum cytoplasm contains frothy gelatinous bubbles called calymma and algal symbionts. Some of the radiolarians emit a bluish flash of light (bioluminescence) to deter predators. The life cycle of some shallow water radiolarians is suggested to be 1–3 months. Binary fission is observed in some species, but it is not established if they reproduce sexually. The soft parts of radiolarians are collectively known as malacoma. The skeleton of a radiolarian, known as a scleracoma, is made of amorphous, opaline silica. An amazing diversity of skeletal structures is found in the living and fossil radiolarians. The skeleton is enclosed within a cytoplasmic sheath called a cytokalymma and not in direct contact with the seawater. The skeleton consists of a porous lattice shell of variable shapes from spherical to spindle and conical. There may be concentric or overlapping lattice shells. The radial elements are hollow to solid spines (attached at one end only), bars (attached at both ends to other elements) and simple spicules. There are three well-recognized divisions of radiolarian, Spumellaria, Nassellaria and Phaeodaria. The orders Spumellaria and Nassellaria are grouped under the Class Polycystinea, and Phaeodaria forms a distinct class. The phaeodarian structures are delicate and formed of 95 % organic

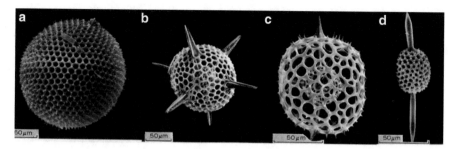

Fig. 7.2 Spumellarian radiolarians characterized by spherical to ellipsoidal shape and perforate wall (reproduced after Kling 1998, with permission © Elsevier)

matter and 5 % silica, and, therefore, have poor fossil records. Anderson (1983) discusses the distinguishing morphological characteristics of these types as follows.

The skeletons of Spumellaria (Fig. 7.2) are characterized by radial symmetry and comprise (1) simple, needle-shaped spicules to complex, symmetrically arranged, triradiate spines distributed in the extracapsulum or clustered around the central capsule, (2) spheroidal to spherical shells that are either single or multiple concentric, enclosing the central capsule, and (3) complex polyhedral skeletons resembling lattices or geodesic structures, reinforced in some groups by radial beams. In the skeletons with concentric shells, the innermost layer is the medullary shell and the outermost shells are cortical shells. The connections between shells and structures are through bars and beams.

The Nassellaria (Fig. 7.3) are characterized by axial symmetry and have three types of skeletons: (1) a tripod located near the base of the central capsule and characterized by three divergent bar-like elements united at a common central point, (2) conical or hat-shaped, complexly perforated shells, and (3) a sagittal ring that reinforces the latticed shell in the medial, sagittal plane. The skeletons of Nassellaria are often very complex and multi-chambered, differentiated into cephalis, thorax and abdomen.

A large, oblate spheroid, depressed in the direction of the main axis, distinguishes the Phaeodaria from the other two types. The skeletons of some Phaeodaria com-

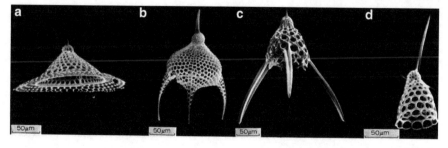

Fig. 7.3 Nassellarian radiolarians characterized by cap-shaped skeletons, small spherical cephalis and one or more post-cephalic chambers (reproduced after Kling 1998, with permission © Elsevier)

prise hollow tubes containing living cytoplasm and organic materials where tubes are joined to one another. Such fragile skeletons get disarticulated before being buried under the sediments. In living forms, the Phaedorians are characterized by distinctive and complex architectures.

Ecology

The radiolarians are exclusively marine. The alveolar complex containing CO_2-saturated water is suggested to be an adaptation for the planktic mode of life. The vertical movement of radiolarians in seawater is supposed to be facilitated by adjustment in volume of the CO_2 in alveoli. The skeletal structures, such as perforated walls, radiating spines and axopods, are further adapted for enhancing buoyancy in the water column. Many radiolarians host dinoflagellates as symbionts and, therefore, dominate the photic zone (<200 m). Some of the radiolarians form colonies of spherical to cylindrical shapes, centimetres to metres in dimension.

The radiolarians are abundantly present in equatorial latitudes, but they also occur in subpolar seas. There are distinct assemblages corresponding to different ocean circulation and water mass characteristics. Seven biogeographic faunal zones are distinguished in the Pacific (Casey 1971). Although several of the species may occur in more than one faunal zone, a simplified distribution of the most abundant species in selected zones is listed below:

Subarctic Transition Fauna: high latitude, north of Arctic or Polar Convergence; *Spongotrochus glacialis, Sethophormis rotula, Pterocanium* sp.

Central Shallow Fauna: intermediate latitude; *Calocyclas amicae, Euchitonia furcata, Eucyrtidium hertwigii.*

Equatorial Fauna: low latitude, bounded by North and South Equatorial Current Systems; *Acrosphaera murrayana, Acrobotrissa cribrosa, Anthocyrtidium cineraria, Lithomelissa monoceras, Tristylospyris scaphipes.*

The statistical analysis of the present-day distribution of radiolarians in the southern hemisphere provides the following transfer functions on the basis of factor loadings for the subtropical (A), Antarctic (B) and sub-Antarctic (C) assemblages:

$$Tw = 15.266\ A + 1.542\ B + 9.984\ C - 1.984,$$

$$Ts = 13.966\ A - 2.904\ B + 10.545\ C + 5.103,$$

where Tw and Ts refer to temperatures in winter (August) and summer (February), respectively. The error in both functions is less than 10 %, which is a fairly good estimate (refer to Lozano and Hays 1976 and Hays et al 1976 for further details). The transfer function can estimate Quaternary seawater temperatures with a fair degree of accuracy.

The radiolarian distributions are also related to upwelling. Radiolarian-based indices are proposed for estimating paleo-upwelling. The Upwelling Radiolarian

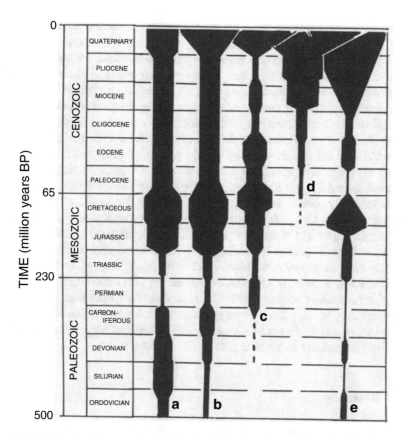

Fig. 7.4 Relative diversity of Spumellaria (*a*) and Nassellaria (*b*) radiolarian through geologic time. The diversities of dinoflagellate (*c*), diatoms (*d*), and tintinnids (*e*) are also shown for comparison (reproduced after Anderson 1983, with permission © Springer Science + Business Media)

Index (URI), based on assemblages in the eastern equatorial Pacific, considered 14 upwelling species, including *Acrosphaera murrayana*, *Eucyrtidium erythromystax*, *Lamprocyrtis nigriniae* and *Pterocorys minythorax* (Haslett 2003).

Geologic History

The fossil record of radiolarians ranges from the Cambrian Era, but it is likely that the skeleton-less forms would have evolved in the Precambrian. Spumellaria dominated in its early history in the Paleozoic, but Nassellaria began to diversify after the Carboniferous (Fig. 7.4). After the Permian–Triassic boundary extinction, the two groups have expanded since the Jurassic. The Mesozoic was the main era of radiolarian radiation. The cause of the radiation is not certain, but one of the reasons was the breakup of the continents and partitioning of the world's oceans that strengthened ocean circulation and the upwelling of nutrient (D'Wever et al. 2003). Because of their high morphological diversity, radiolarians are good tracers of evolutionary history. An intensive study of evolutionary trends of Cretaceous radiolarians reveals that Spumellaria and Nassellaria respond differently to ecological stress. The spumellarians are much more extinction resistant during episodes of critical environmental deterioration, but nassellarians possess a relatively higher evolutionary rate than spumellarians during periods of radiation (O'Dogherty and Guex 2002).

The diversity of radiolarians was stable for the greater part of the Cenozoic, but a distinct change in the robustness of the test occurred after the Eocene. The Oligocene was marked by the demise of many thickly silicified Paleogene radiolarians. The average weight of the radiolarian test has decreased four times since the Eocene epoch (Racki and Cordey 2000). Variations in temperature, nutrient and intensity of light in the water column are the significant factors in the evolution of radiolarians and other plankton. Competition with the co-evolving group also played a major role. The diminishing weight of the test is attributed to competitive pressure for dissolved silica in seawater triggered by the explosive radiation of other siliceous plankton, including diatoms and silicoflagellates. The radiolarians adapted by decreasing the wall's thickness, increasing pore size and decreasing the width of the bars. A visible expansion in diversity occurred in the Quaternary and the nassellarians, in particular, diversified more in this period. There is a fair similarity between the diversity trends of radiolarians and dinoflagellates, possibly indicating a trophic relationship or symbiotic interdependence (Anderson 1983).

7.2   Diatoms

A diatom is a single-celled, free-floating or attached golden-brown algae belonging to the Class Bacillariophyceae of the Phylum Chrysophyta. It lives in the photic zone of marine water, brackish water, freshwater, ice and in soils. All living diatoms have a silicified cell wall except for some diatoms living as symbionts in dinoflagellates and foraminifera. Its skeleton is called a frustule and it is made of opaline silica. Diatoms are primary producers and contribute more than radiolarians to the biogenic silica in the ocean. The sediments of high productivity areas, particularly in high latitudes and coastal upwelling, are exceedingly rich in diatoms, often forming siliceous rocks called diatomites. The diatomites have numerous industrial applications, including filtration and pharmaceuticals.

Morphology

The diatom cell contains a nucleus and photosensitive chromatophore in the protoplasmic mass. The cytoplasm is highly vacuolated, which probably aids in buoyancy. The diatom life cycle consists of both asexual and sexual phases. The opaline silica frustule secreted by the diatom consists of two valves of petri dish shape. The valves are 10–100 μm in length, and one valve fits over the other. The larger valve

Fig. 7.5 Schematic
diagrams of morphology
and main features of
centric and pinnate diatoms
(redrawn after Hasle and
Syvertsen 1996, with
permission © Elsevier)

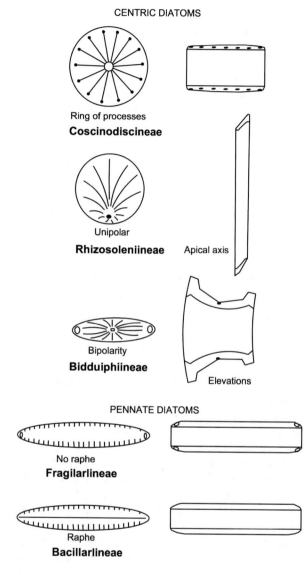

CENTRIC DIATOMS

Ring of processes
Coscinodiscineae

Unipolar
Rhizosoleniineae   Apical axis

Bipolarity
Bidduiphiineae

Elevations

PENNATE DIATOMS

No raphe
Fragilarlineae

Raphe
Bacillarlineae

is the epitheca and the smaller one is the hypotheca. A thin circular band called a
girdle connects the valves (Fig. 7.5). Morphologically, there are two kinds of diatom
and, correspondingly, two orders of diatom, the Centrales and the Pennales, respectively. The centric forms are circular, triangular or oblong and their surface markings radiate from a central area. The pennate forms are elongate and the surface
markings are at right angles to the long axis. Pores called punctae cover a large part
of the surface area of the valves. Striae refer to the arrangement of punctae in lines.
Many diatoms have honeycomb structures termed areolae. The areolae consist of

vertical-sided chambers within the valve wall and may be covered by a sieve plate. Some diatoms are characterized by clear areas of structure-less silica termed hyaline areas. Some pennate forms have a V-shaped slit called raphe. The raphe is distinguished by rows of punctae on either side that facilitate the flow of mucous. It allows for a creeping motion and helps diatoms to glide over the sediments. The girdles, symmetry or asymmetry of the frustules, raphe, direction of terminal fissures, form and presence of axial and central areas and valve structures are the morphological characters used in identification and classification of diatoms.

Ecology

Diatoms live in both marine and freshwaters, each with their own distinctive assemblage. The representative diatoms of the different aquatic environments are as follows:

Freshwater lakes: *Fragilaria, Cyclotella, Rhizosolenia, Asterionella*.
River: *Achnanthes, Cocconeis, Cymbella*.
Marine: *Triceratium, Campylodiscus, Chaetoceros, Coscinodiscus*.

The pH of the ambient water is an important chemical factor influencing diatom distribution and, accordingly, the inhabiting species are categorized as acidobiontic (pH <7), acidophilus (pH ~7) and alkalibiontic (only in alkaline waters). In the absence of the commonly used calcareous microfossil foraminifera from the freshwater lacustrine sediments, diatoms are most useful in paleolimnology for climate interpretation. A study on the relationship between present-day diatoms and the environmental variables of lakes located at different altitudes has resulted in a transfer function for estimation of temperature for lacustrine sediments (Vyverman and Sabbe 1995). It can potentially estimate lake-water temperatures to an accuracy of about 3 °C.

In marine environments, diatoms prefer high productivity and upwelling areas, including those of Antarctic divergence and off the coast of Peru. In the upwelling regions of the NW Indian Ocean (off Somalia), 80 % of the deposition of *Thalassionema nitzschioides* and *Chaetoceras* occurred during the upwelling periods (Koning et al. 2001) and, therefore, are good indicator taxa of paleo-upwelling intensity. The abundance of frustules in the sediments of these regions can exceed 100 million per gram of sediments. There are also seasonal variations in diatom abundance, and during the spring and late summer blooms in high latitudes, the number can be as high as 1000 million cells per m^3 of water. The intensity of light, temperature and salinity are important controls on the distribution of diatoms. The diatoms have, however, successfully colonized the polar regions due to their adaptability to low temperatures, periodically low light intensity and high salinity. The biogeography of the modern planktic diatoms suggests that some species are confined to cold waters, some occur only in warm waters, and others are cosmopolitan in distribution. It is also observed that, morphologically and taxonomically, closely related species may have completely different ecological distribution. The following distribution of different species of two genera, *Nitzschia* and *Thalassiosira*, is a good example of how reliable identification at the species level is a key to past seawater temperature reconstruction (see Hasle 1976 for complete assemblage):

Coldwater: Northern hemisphere	*N. grunowii, N. paaschei, N. seriata, T. constricta, T. hyalina*
Bipolar	*N. cylindrus, T. antarctica*
Southern hemisphere	*N. curtata, N. ritscheri, N. saparanda, T. gracilis, T. ritscheri*
Warm water	*N. marina, N. subpacifica, T. lineate, T. minuscula*
Cosmopolitan	*N. bicapitata, N. pungens, T. eccentric, T. profunda*

Several of the species in the indicator assemblages can be traced back to the Miocene and, therefore, diatoms are one of the most useful groups of microfossils in paleoceanography for reconstructing paleotemperature and paleoproductivity. They are also important in biostratigraphic zonations of deep-sea sequences, particularly in high latitude areas where other microfossils are poorly preserved.

The high productivity of diatoms in certain areas leads to the formation of diatomites. Three main areas of high diatom productivity in the present time include the subarctic waters of the northern hemisphere, the sub-Antarctic waters of the southern hemisphere and the equatorial belt of the Indian and Pacific Oceans. In addition to high diatom productivity, low terrigenous influx and high solubility of calcium carbonate are essential for the formation of diatomites. It is important to know that the diatom frustules are prone to dissolution in silica-undersaturated modern seawater, and it is estimated that only a small fraction of the diatom frustules reach the seafloor. The taphonomic biases should, therefore, be given due consideration while interpreting the stratigraphic records.

Geologic History

The diatoms are mostly known from the Cretaceous and younger strata, although their earliest record is from the Jurassic. The oldest record is of two species of *Pyxidicula* from inner-shore marine sediments of the Toarcian Stage (Early Jurassic). It implies that the earliest diatoms were marine and invaded freshwaters only in the later part of the Mesozoic. The early diatoms were centric forms and were joined by pennate types in the late Cretaceous and raphid diatoms in the Paleocene. The molecular data also supports the fossil records indicating that pennate types evolved from the centric forms. Considerable morphological diversification took place in the Cretaceous. Like the other groups of marine biota, diatoms also suffered extinction at the K–T boundary, but the impact was much less severe. Several morphological and taxonomic changes occurred in the Cenozoic. There was a marked decline in many old genera and families, and the appearance of several new ones. Some of the salient points of diatom evolution in the Cenozoic are summarized below (see Sims et al. 2006 for details):

1. Diatoms moved from the shallow shelf to deeper oceanic waters.
2. Appearance of the raphe system provided motility to the group and, as a result, diatoms expanded to a wide range of benthic habitats.
3. Explosive diversification occurred in the Middle and Late Eocene. The robust centric genera, dominating the Paleocene and Eocene, were gradually replaced by smaller and more delicate genera in the colder waters of the Early Oligocene. The high and low latitude assemblages began to differentiate at this time.
4. Small and delicate genera such as *Chaetoceras* and *Skeletonema* dominate the diatom blooms in the present-day coastal upwelling zones. This also implies that diatoms have adapted to the silica-undersaturated present-day ocean by economizing silica utilization in building their skeletons.

7.3   Silicoflagellates

The silicoflagellates are opal-secreting unicellular phytoplankton, ranging in size from 20 to 100 μm and referred to as Chrysophyceae or golden algae. They appeared for the first time in the early Cretaceous. Silicoflagellates are exclusively marine and live in the photic zone. They form only a minor part of the present-day phytoplankton communities, but constitute an important group of microfossils in paleoceanography and paleoclimatology.

Morphology

The cell of the silicoflagellates contains a nucleus and chromatophores for photosynthesis. The protoplasm is enclosed within a siliceous skeleton. The pseudopodia and a flagellum extend outside the skeleton. Silicoflagellates are autotrophic and obtain their food through photosynthesis. The skeleton is made of opaline silica and morphologically varies from a ring-like structure with spines to elongated and dome-shaped structures. A basal ring is present in all forms on which complex structures develop. The basal ring may be circular, trigonal, quadrate or hexagonal, with spines or knobs at the corners (Fig. 7.6). A dome-shaped apical structure is present above the basal ring. The apical structure consists of elements called struts attached to the basal ring. The morphologic variations in the skeletal structure recognize some major groups in silicoflagellates that are named after the typical genus of the group. For example, the three-sided forms are *Corbisema*, forms with four sides and up and without apical rings are *Dictyocha* and forms with apical rings are *Distephanus*.

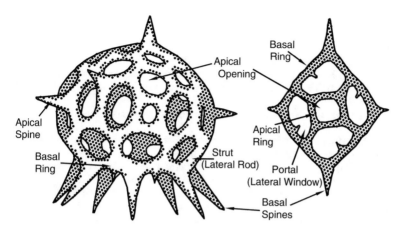

Fig. 7.6 General morphology of silicoflagellate skeletons

Ecology

The distribution of silicoflagellates is controlled by water mass characteristics, including temperature and salinity. As mentioned, they occur in the photic zone and there are distinct latitudinal variations in the assemblage. For example, *Dictyocha* occurs in low to mid-latitudes and *Distephanus* in high latitudes. The ratios of the two genera are used to infer climatic fluctuation. Both nutrients and silica are important for silicoflagellates and, therefore, they occur abundantly in upwelling areas. Takahashi and Blackwelder (1992) found the following quantitative relationship between salinity and ratios of the two silicoflagellates species:

$$Dm/Ds = 6.81 \times 10^{-36} \times 10^{0.98S},$$

where Dm and Ds are the abundances of *Dictyocha messanensis* and *Distephanus speculum*, respectively, and S is salinity.

Geologic History

Silicoflagellates evolved in the Cretaceous. *Lyramula* and *Vallacerta* are the earliest genera, and they disappeared at or near the K–T boundary. *Dictyocha* and *Corbisema* crossed the boundary and were joined by *Naviculopsis* and *Mesocena* in the Paleocene. Only three genera have been present since the Pleistocene, *Distephanus*, *Dictyocha* and *Octactis*. Due to slow evolutionary rates, the taxa are long ranging and, therefore, the silicoflagellate biozones are comparatively few in numbers and long ranging. They are useful only for local or regional correlations and for high latitudes.

References

Anderson OR (1983) Radiolaria. Springer, New York

Casey RE (1971) Radiolarians as indicators of past and present water-masses. In: Funnell BM, Riedel WR (eds) The micropalaeontology of oceans. Cambridge University Press, Cambridge, pp 331–339

Hasle GR (1976) The biogeography of some marine planktonic diatoms. Deep-Sea Res 23:319–338

Hasle GR, Syvertsen EE (1996) Marine diatoms. In: Tomas CR (ed) Identifying marine diatoms and dinoflagellates. Academic, San Diego, pp 5–385

Haslett SK (2003) Upwelling related distribution patterns of radiolarians in Holocene surface sediments of the eastern equatorial Pacific. Rev Esp Micropaleontol 35:365–381

Hays JD, Lozano JA, Shackletin N, Irving G (1976) Reconstruction of the Atlantic and western Indian Ocean sectors of the 18000 B.P. Antarctic Ocean. Geol Soc Am Memoir 145:337–374

Kling SA (1998) Radiolaria. In: Haq BU, Boersma A (eds) Introduction to marine micropaleontology. Elsevier, Singapore, pp 203–244

Koning E, van Iperen JM, van Raaphorst W, Helder W, Brummer GJA, van Weering TCE (2001) Selective preservation of upwelling indicating diatoms in sediments off Somalia, NW Indian Ocean. Deep Sea Res I 48:2473–2495

Lozano JA, Hays JD (1976) Relationship of radiolarian assemblages to sediment types and physical oceanography in the Atlantic and western Indian Ocean sectors of the Antarctic Ocean. Geol Soc Am Memoir 154:303–336

O'Dogherty L, Guex J (2002) Rates and pattern of evolution among Cretaceous radiolarians: relations with global paleoceanographic events. Micropaleontology 48(Suppl 1):1–22

Racki G, Cordey F (2000) Radiolarian palaeoecology and radiolarites: is the present the key to the past? Earth Sci Rev 52:83–120

Sharma V, Daneshian J (1998) Radiolaria as tracers of ocean-climate history. Curr Sci 75:893–897

Sims PA, Mann DG, Medlin LK (2006) Evolution of the diatoms: insights from fossil, biological and molecular data. Phycologia 45:361–402

Takahashi K, Blackwelder PL (1992) The spatial distribution of silicoflagellates in the region of the Gulf Stream warm-core ring 82B: application to water mass tracer studies. Deep Sea Res 39(Suppl 1):327–347

Vyverman W, Sabbe K (1995) Diatom-temperature transfer functions based on the altitudinal zonation of diatom assemblages in Papua New Guinea: a possible tool in reconstruction of regional palaeoclimatic changes. J Paleolimnol 13:65–77

D'Wever P, O'Dogherty L, Coridroit M et al (2003) Diversity of radiolarian families through time. Bull Soc Geol Fr 174:453–469

Further Reading

Anderson OR (1983) Radiolaria. Springer, New York

Lipps JH (1993) Fossil prokaryotes and protists. Blackwell, Boston

Chapter 8
Phosphatic Microfossils

8.1 Conodonts

Conodonts are an extinct group of primitive jawless vertebrates belonging to the Phylum Chordata. They were largely soft-bodied organisms except for the presence of a feeding apparatus comprised of tooth-like elements that are preserved as fossils. The elements range in size from 0.25 to 2 mm and are calcium phosphate in composition. The biological affinity of conodonts was debated for a long time, until a well-preserved complete conodont animal was discovered (Briggs et al. 1983). Conodonts range in age from the Late Cambrian to the Late Triassic. In spite of their until-recently-unknown affinity, they have been one of the most significant and widely used groups of fossils in Paleozoic stratigraphy. For instance, the discovery of conodonts in the so-called unfossiliferous formations of the Lesser Himalayas enabled resolution of the long-standing controversy over the age of these sequences (Azmi et al. 1981). The alteration in the colour of the conodont elements, expressed as the Conodont Colour Alteration Index (CAI), is an indicator of thermal maturity and, hence, useful in basin analysis for hydrocarbon exploration.

Morphology

The anatomy of the conodont animal is known mainly due to the very well-preserved species *Clydagnathus windsorensis* from the Carboniferous of Scotland and *Promissum pulchrum* from the Ordovician rocks of South Africa. Aldridge et al. (1993) provide anatomical details of *Clydagnathus*. The animals were 20–55 mm in length, with a short head characterized by two lobate structures, marking the positions of large eyes. In the ventral and immediately posterior to the eyes were a

© Springer International Publishing Switzerland 2016
P.K. Saraswati, M.S. Srinivasan, *Micropaleontology*,
DOI 10.1007/978-3-319-14574-7_8

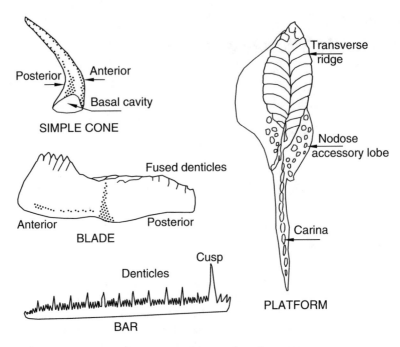

Fig. 8.1 The major morphological features of the conodont elements

feeding apparatus, formed of phosphatic, tooth-like elements. The trunk of most specimens displayed the notochord as a pair of axial lines. The tail was present with ray-supported fins representing a caudal fin, similar to hagfish and lampreys, the most primitive living vertebrates (Sweet and Donoghue 2001). Briggs (1992) discussed the biological affinity of the conodonts as follows. The growth pattern and arrangement of crystallites within the lamellae in the elements of many conodonts are typical of the structure of enamel in vertebrate teeth. The areas of opaque "white matter" occurring in crowns include features identical to those in cellular bone. The evidence of soft tissues is indicative of two possible affinities: (1) a sister group of Myxinoidea (hagfishes) or (2) a primitive sister group of the higher craniates (other than Myxinoidea).

The use of the term "conodont" is generally accepted to mean the complete animal, while the term "element" refers to the mineralized skeletal structure. The main components of the conodont elements are the "crown" (the part that looks like a tooth) and the "basal body" (under the crown). The largest cone-shaped structure in the crown is the "cusp". The conodont elements are of the following major types (Fig. 8.1):

Simple cones: These are formed of a single tapered cusp, which may be smooth or have costae. Cones are pointed at the oral end and expand at the base with the basal cavity. The cones are of ageniculate type if the concave edge of the base joins the cusp at an acute angle and of a rastrate type if the denticles developed

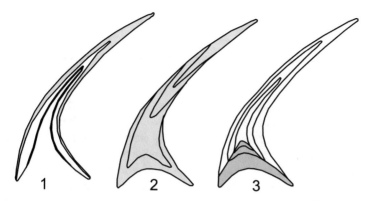

Fig. 8.2 Bengtson's histologic groups of conodonts: (*1*) Protoconodont (*2*) Paraconodont and (*3*) Euconodont

along the concave edge of the cusp. The coniform elements are important markers of the Cambrian and Ordovician and became extinct in the Devonian.

Bar-type (or ramiform) elements: These are laterally compressed, thin bars with discrete denticles. They developed from simple cones and are long-ranging forms and, hence, of little biostratigraphic value.

Blade-type (or ramiform) elements: These are elongate, laterally compressed elements with rows of denticles that are fused except at their tips. They are useful for biostratigraphy of the Silurian to the Triassic.

Platform-type (or pectiniform) elements: These are broad flanges of bar or blade types that probably evolved into plates with highly differentiated elements comprising ridges, sulcus and nodes on one side and a serrate, denticulate blade on the other. Many platform-type elements are index fossils for the Ordovician to the Triassic.

A scheme of anatomical notation and orientation has been devised to bring consistency to the use of terms for orientation of the conodont elements. This is according to the topological relationship between elements with reference to the principal axes of the body. It is known as the P_n–S_n scheme, in which the elements are expressed in the form of letters with subscripts (refer to Purnell et al. 2000 for a detailed discussion).

There were a variety of simple, tiny, cone-shaped fossils in the Proterozoic and early Paleozoic that were organic or weakly phosphatic. Not all were conodonts. Three histologic groups are recognized (Fig. 8.2; Bengtson 1976): paraconodont (single layer of tissue), euconodont (two layers of tissue) and protoconodont (three layers of tissue). Electron microscopic study has revealed the three layers of protoconodont as being a thin outer organic cover, a thick, laminated, organo-phosphatic middle layer and a thin, faintly laminated, organic inner layer (Szaniawski 1983). The protoconodonts and paraconodonts are grouped together as Paraconodontida and were previously regarded as conodonts. Both groups are now excluded from the true conodonts and the Conodonta includes only the euconodonts.

Paleoecology

The life history and ecology of the conodont animals are known only indirectly from the associated fauna and facies. It is inferred that the conodonts were exclusively marine but occupied a range of habitats. The facies-independent occurrences of many conodonts, from bedded cherts deposited below the CCD to shallow marine tropical to subtropical limestone, indicate that they were nektonic or pelagic animals. The well-known records of the Granton Beds of Scotland are from shallow marine, higher salinity, quiet water environments, unlike the occurrences of other conodonts in open marine facies. There are several reports of conodonts from deep-water black shales where benthic fauna are absent. These observations support the view that conodonts were nektonic if not pelagic. It may not be possible to define the trophic characters of the conodonts with any certainty, but the results of functional morphologic analysis of the apparatus and the elements are interesting in this context. Zhuravlev (2007) reconstructed five major functional types of conodont elements: grasping—holding, filtering, cutting, crushing and grinding (Fig. 8.3). These types have the potential to interpret the trophic characters as to whether the conodonts were suspension feeders, detritus feeders or predators. The provincialism in conodont assemblages of certain times suggests that temperature was an important control factor in their distribution. The oxygen isotopic data (Luz et al. 1984) indicates that conodonts lived in warmer seas and could tolerate temperatures in excess of 40 °C.

Geologic History

The true conodonts first appeared in the Late Cambrian. It is stated that euconodonts evolved from paraconodonts by acquiring a crown, a dense tissue of apatite secreted over the outer surface of the paraconodont cusp (Bengtson 1983). The oldest representatives are referred to as the Order Proconodontida. They are characterized by coniform elements, subsymmetrical or oval in cross section. The initial simple apparatus progressively differentiated into several morphologic types and other orders were established by the time the conodonts reached the acme of their diversity in the Arenig Stage (Early Ordovician). The orders Protopanderodontida, Panderodontida and Prioniodontida appeared in this phase of evolution. The Protopanderodontida also had coniform elements in which the basal cavity was short in relation to the cusp. The coniform Panderodontida had lateral furrows. The order Prioniodontida had platform-type elements, evolved in the Early Ordovician and became extinct by the end of the Devonian. There was a general decline in diversity of conodonts after their acme in the Early Ordovician that continued through the Silurian and into the early Devonian. Diversity continued to decrease and, eventually, conodonts became extinct in the Late Triassic. The two orders that survived until the Triassic were Prioniodinida and Ozarkodinida. The paleobiogeographic distribution of conodonts suggests that, in the Cambrian and much of the Ordovician, the high latitude fauna were quite different from the low latitude fauna.

Fig. 8.3 The major functional types of conodont elements (from *top to bottom*)—grasping, filtering, cutting, crushing and grinding (reproduced after Zhuravlev 2007, with permission © Springer Science + Business Media)

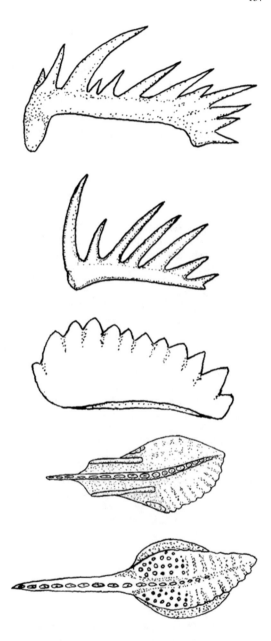

The differences began to diminish by the end of the Ordovician, and by the Silurian, the two faunas were quite similar to each other (Sweet 1985). The end of provinciality in the fauna could have been the return of an equitable climate. The Ordovician to Pennsylvanian conodonts of North America were analysed for oxygen isotopic composition to estimate seawater temperatures. The $\delta^{18}O$ values of the conodont apatite suggest a temperature maximum of 43 °C in the Late Devonian that declined considerably to less than 30 °C by the end of the Pennsylvanian (Luz et al. 1984).

8.2 Other Phosphatic Microfossils

There are several other minor groups of phosphatic microfossils in the Cambrian and Ordovician. These included Hyolithelminthes, horny brachiopods, phosphatic ostracode-like arthropods and bone fragments. The Hyolithelminthes were conical phosphatic tubes 5–15 mm in length and consisted of the common genera *Hyolithellus* and *Lapworthella* (Müller 1978).

References

Aldridge RJ, Briggs DEG, Smith MP, Clarkson ENK, Clark NDL (1993) The anatomy of conodonts. Philos Trans R Soc Lond Biol Sci 340(1294):405–421
Azmi RJ, Joshi MN, Juyal KP (1981) Discovery of the Cambro-Ordovician conodonts from Mussoorie Tal Phosphorite: its significance in correlation of the Lesser Himalaya. In: Sinha AK (ed) Contemporary scientific researches in Himalayas, 1st edn. Bishen Singh Mahindra Pal Singh, Dehradun, India, pp 245–250
Bengtson S (1976) The structure of some Middle Cambrian conodonts and the early evolution of conodont structure and function. Lethaia 9:185–206
Bengtson S (1983) The early history of the Conodonta. Fossils Strata 15:5–19
Briggs DEG (1992) Conodonts: a major extinct group added to the vertebrates. Science 256(5061):1285–1286
Briggs DEG, Clarkson ENK, Aldridge RJ (1983) The conodont animal. Lethaia 16:1–14
Luz B, Kolodny Y, Kovach J (1984) Oxygen isotope variations in phosphate of biogenic apatites, III. Conodonts. Earth Planet Sci Lett 69:255–262
Müller KJ (1978) Conodonts and other phosphatic microfossils. In: Haq BU, Boersma A (eds) Introduction to marine micropaleontology, IIth edn. Elsevier, New York, pp 277–291
Purnell MA, Donoghue PCJ, Aldridge RJ (2000) Orientation and anatomical notation in conodonts. J Paleontol 74:113–122
Sweet WC (1985) Conodonts: those fascinating little whatzits. J Paleontol 59:485–494
Sweet WC, Donoghue PCJ (2001) Conodonts: past, present, future. J Paleontol 75:1174–1184
Szaniawski H (1983) Structure of protoconodont elements. Fossils Strata 15:21–27
Zhuravlev AV (2007) Morphofunctional analysis of Late Paleozoic conodont elements and apparatuses. Paleontol J 41:549–557

Further Reading

Armstrong HA, Brasier MD (2005) Microfossils, IIth edn. Blackwell, Oxford
Müller KJ (1978) Conodonts and other phosphatic microfossils. In: Haq BU, Boersma A (eds) Introduction to marine micropaleontology, IIth edn. Elsevier, New York, pp 277–291

Chapter 9
Organic-Walled Microfossils

9.1 Dinoflagellates

Dinoflagellates are single-celled organisms, belonging to Phylum Dinozoa and Subphylum Dinoflagellata. Dinoflagellates are both autotrophic and heterotrophic, and several of them occur as symbionts in corals, foraminifera and a number of other invertebrates. They are predominantly marine, but also occur in brackish and fresh waters. Some dinoflagellates are bioluminescent and cause sparkling in seawater due to the chemical reaction of an enzyme. Dinoflagellates are among the major primary producers in the present-day oceans, along with diatoms and coccolithophores. The free-swimming (motile) cells are abundant, but the resting cysts (the dinocysts) are resistant and occur in fossil records. The distribution of present-day dinoflagellates is limited by temperature, salinity and nutrient of the seawater and, therefore, fossil dinocysts are potential indicators of past environments. Although dinoflagellates are reported in Silurian strata, the earliest definitive record is from the Late Triassic. Due to preservation of the cyst stage alone, the fossil record is incomplete and biased. Some dinoflagellates have produced fossilizable remains while many have not, and those groups that have produced fossilizable cysts have done so inconsistently (Evitt 1985). This has several implications for the study of dinoflagellates, including the true assemblages and population size in the geological records, and, therefore, a cautious approach to the use of dinoflagellate data in paleoenvironmental interpretation and biostratigraphy is necessitated.

Morphology

The dinoflagellate cell contains a nucleus, an endoplasmic reticulum, a Golgi apparatus, and mitochondria. Two different cell types are distinguished. The naked or unarmoured type is fragile. The armoured dinoflagellate has cellulose or polysaccharides

© Springer International Publishing Switzerland 2016
P.K. Saraswati, M.S. Srinivasan, *Micropaleontology*,
DOI 10.1007/978-3-319-14574-7_9

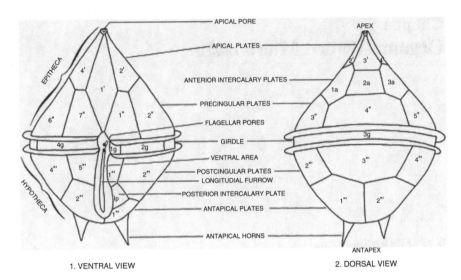

Fig. 9.1 Principal thecal structures and symbols used to identify individual plates of dinoflagellates (reproduced after Evitt 1961, with permission © Micropaleontology Press)

within each vesicle, called theca, making the wall more rigid. The wall of the theca is commonly divided into a number of polygonal areas, termed plates. The pattern of arrangement of the plates is called the tabulation. In some forms, the theca is impregnated with silica or it may be entirely siliceous or calcareous. A series of plates encircle each cell. In the motile stage, the living dinoflagellates propel themselves through the water with two flagella. The furrows on the cell surface that bear the flagella are either equatorially positioned (known as cingulum) or longitudinally placed (known as sulcus). The half of the cell anterior to the cingulum is called the epitheca and the other half posterior to it is the hypotheca. The plates are given different names and each plate is given a symbol based on its position, known as the Kofoidian symbol. From the apex to the antapex, these are apical ($'$), anterior intercalary (a), and precingular series ($''$) of the epitheca, the cingular series (c) composing the cingulum, and the postcingular ($'''$), posterior intercalary (p) and antapical series ($''''$) of the hypotheca (Fig. 9.1; refer Evitt 1985 for details). Many dinoflagellates break open to release their contents. Such ruptured thecae are found in both living and fossil dinoflagellates. These may rupture (1) along the girdle, (2) along the line of sutures between the apical and precingular series of plates, or (3) by release of a single plate of epitheca. The structures formed due to rupturing of plates are known as "archeopyle". Three types of archeopyle are distinguished: apical, intercalary and precingular (Fig. 9.2).

The predominant mode of reproduction in the dinoflagellate is asexual, in which a cell divides into two halves by binary fission. Sexual reproduction is known in very few forms. During reproduction, protoplasm may first become enclosed in a

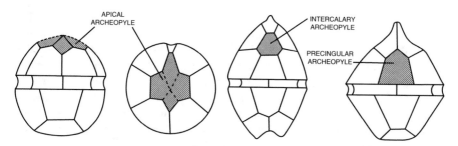

Fig. 9.2 Types of archeopyle in dinoflagellate cysts: apical archeopyle (*first* and *second* from *left*), intercalary archeopyle (*third*) and precingular archeopyle (*fourth*) (reproduced after Evitt 1961, with permission © Micropaleontology Press)

toughened wall, called a cyst. However, a cyst may also form to protect the dinoflagellate against adverse environmental conditions, or as a refuge from predation and nutrient depletion (Sarjeant et al. 1987). Usually, the encysted dinoflagellates, due to their resistant surfaces, are preserved as fossils. The cyst wall, called the phragma, is made of organic material called dinosporin. It may consist of single or multiple layers. The surface of the cyst may be smooth or may have processes, including granules, ridges, spines and horns. Different types of dinoflagellate cysts are distinguished. The proximate cyst resembles a theca in size and shape and it reflects tabulation, cingulum and sulcus. The chorate cyst usually does not have any trace of a reflected cingulum or sulcus. The proximochordate cyst is an intermediate type between the two. A fourth type, called cavate cyst, is also distinguished by the fact that the two wall layers are markedly separated. It should be mentioned that the morphology of the dinoflagellate cysts is entirely or largely a reflection of the internal morphology of the motile dinoflagellates (Sarjeant et al. 1987).

Ecology

Dinoflagellates live in all aquatic environments from normal marine to brackish and freshwaters. Although as a group they are tolerant of variable salinities, none of the species moves between marine and fresh waters. Most species of *Ceratium* are stenohaline, while *Gymnodinium* and *Peridinium* are euryhaline, able to tolerate a range from freshwater to normal marine water. Several species develop morphological aberrations in response to low salinity. *Protoceratium reticulatum*, *Spiriferites* and *Lingulodinium polyedrum* have shorter and thicker processes in low salinity fjords (Ellegaard 2000). In the Cenozoic, members of the *Homotryblium* complex were indicators of restricted settings with increased salinity (Sluijs et al. 2005).

The motile stage of the autotrophic forms lives in the photic zone. Some species of *Ceratium* and *Peridinium* are shallow dwellers, while the species of *Heterodinium* and *Triposolenia* live in deeper water. Autotrophic forms live in areas of upwelling rich in nutrients. In a study on *Protoperidinium* cysts in the Arabian Sea, it was found that there is a strong correlation between its absolute abundance and other upwelling proxies (Reichart and Brinkhuis 2003). The *Protoperidinium* concentration, therefore, can be used to reconstruct productivity. The dinoflagellates of mid to high latitude environments suggest that some species, such as *Operculodinium centrocarpum*, are ubiquitous, but most taxa have a narrow range of tolerance for temperature, salinity and/or seasonal duration of ice cover. *Impagidinium strialatum* and *Impagidinium aculeatum* are thermophilous and *Impagidinium pallidum* is a cold water species (de Vernal et al. 1997). Several authors have defined distinct regions in the modern oceans based on species of *Ceratium* that closely correspond to hydrographic conditions. The warm and cold waters are also distinguished by the morphotypes of *Ceratium*. The *Ceratium* of tropical waters have longer horns and thinner thecae compared with those from cold waters (Williams 1971).

Geologic History

The Late Silurian species *Arpylorus antiquus* is believed to be the oldest record of a dinoflagellate. Several authors, however, do not consider it a true dinoflagellate. There is no fossil record of dinoflagellates for the next 200 million years, spanning the rest of the Paleozoic and the Early Triassic. *Rhaetogonyaulax* from the Rhaetian (Late Triassic) is the first unequivocal fossil record of the dinoflagellate. Species diversity began to increase after the early Jurassic and continued into the Cretaceous. The tabulate proximate form with apical archaeopyle (e.g. *Dapcodinium*) and the first chorate cysts (e.g. species belonging to *Polysphaeridium*) had appeared by the Early Jurassic. The dinoflagellate genera diversified in the rest of the Jurassic. Proximochorate and chorate cysts predominated during the Turonian to the Campanian Stages (Late Cretaceous). With the genera *Dynogymnium*, *Prolixosphaeridium* and *Systematophora* appearing at or near the end of the Maastrichtian (latest Cretaceous), cavate cysts became predominant and the dinoflagellates had acquired the Cenozoic aspect by this time (Sarjeant 1974). Dinoflagellates were not as affected by the K/T extinction event as the other phytoplankton. The diversity peaks of dinoflagellates occurred in the Albian and the Maastrichtian in the Cretaceous and Early Eocene (Sluijs et al. 2005). Some of the common taxa in tropical India in the Eocene included *Operculodinium*, *Glaphyrocysta*, *Homotryblium* and *Thalassiphora* (Fig. 9.3). The number of species has declined steadily since the Early Eocene. The genus *Apectodinium* bloomed during the Paleocene–Eocene Thermal Maximum (PETM). Globally high sea surface temperatures, terrestrial discharge and high primary productivity in marginal seas favoured its acme (Crouch et al. 2003). The freshwater dinoflagellates are known only from the Cenozoic. The dinocysts of freshwater *Ceratium* are not potentially fossilizable.

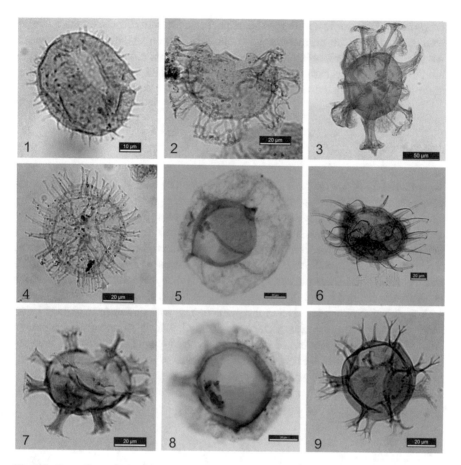

Fig. 9.3 Some Eocene Dinoflagellate cysts from India: *Operculodinium* (*1, 6*) *Glaphyrocysta* (*2*) *Cordosphaeridium* (*3*) *Polysphaeridium* (*4*) *Thalassiphora* (*5*) *Homotryblium* (*7*) *Thalassiphora* (*8*) *Achomosphaera* (*9*) (courtesy Jyoti Sharma)

9.2 Acritarch

The acritarchs are organic-walled and unicellular *incertae sedis* microfossils (that is, of uncertain affinity). A majority of them are believed to be abandoned organic envelopes of protistan phytoplankton, including prasinophyte algae, chlorophyte cysts, dinoflagellate cysts or similar groups (Chuanming et al. 2001, and references therein). Biomarker studies of acritarchs in recent years have indicated their polyphyletic nature. The presence of dinosteranes and other steranes associated with dinoflagellates in selected Early Cambrian acritarchs suggests that some of them are related to dinoflagellates (Talyzina et al. 2000). The Proterozoic acritarch *Chuaria circularis* reveals a predominance of *n*-aliphatic pyrolysates and, hence, their algal affinity (Dutta et al. 2006). They were the main phytoplankton in the Late

Precambrian, Paleozoic and Early Mesozoic. In the Mesozoic and Cenozoic marine
sequences, acritarchs commonly co-occur with dinoflagellate cysts. They are
recorded in near shore, offshore and deep basinal marine sequences. Acritarchs are
used widely in hydrocarbon exploration for biostratigraphy, paleoenvironmental
interpretation and the thermal history of the sedimentary basins.

Morphology

The main body of acritarchs is known as a vesicle. The wall of the vesicle is a complex of polymers, called sporopollenin. The vesicles have wing-like projections
called flanges and large projections known as processes (Fig. 9.4). The shape of the
vesicle is highly variable. Most acritarchs were probably spheroidal or ovoid, as
indicated by the well-preserved specimens recorded in recent years. They, however,
typically occur in two-dimensional compressed forms in fossils. Some acritarchs
show openings for excystment, similar to the archaeopyle in dinocysts. The excystment structure varies from a partial rupture in the vesicle to a circular opening
with a plug-like operculum called the pylome. The wall of the vesicles may be of
single or double layers and smooth or ornamented. This is an important diagnostic
character in the identification of acritarchs. Several types of wall structure are recognized. The micrhystridian type, exemplified by *Micrhystridium*, has a single, thin,
homogeneous wall. The tasmanitid type, of the *Tasmanites*, is characterized by a
thick, laminated wall with narrow radial pores. And finally, the visbysphaerid type,

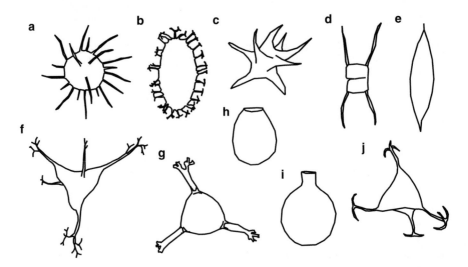

Fig. 9.4 Basic shapes of acritarchs: spheroidal (*A*), ellipsoidal (*B*), stellate (*C*), rectangular (*D*),
fusiform (*E*), triquitrate (*F*), triapsidate (*G*), ovoid (*H*), flask-shaped (*I*), tetrahedral (*J*) (redrawn
after Strother 2002, with permission © American Association of Stratigraphic Palynologists
Foundation)

of the *Visbysphaera*, is double walled (Dorning 2005). Some authors consider *Tasmanites* to belong to the prasinophytes. Acritarchs are broadly divided as (1) those without processes or flanges, (2) those with flanges but without processes, and (3) those with processes but with or without flanges. Each of the three divisions further includes one or more sub-groups. The processes on an individual may be homomorphic (all of similar type) or heteromorphic (of more than one type). Three-dimensionally preserved phosphatized acritarchs from the terminal Proterozoic Doushantuo Formation of China give new insight into the morphology of the group (Fig. 9.5). The vesicles of the two new genera *Bacatisphaera* and *Castaneasphaera* in the formation reach the giant size of 450 μm and are characterized by radially symmetrical, spinose morphology (Chuanming et al. 2001).

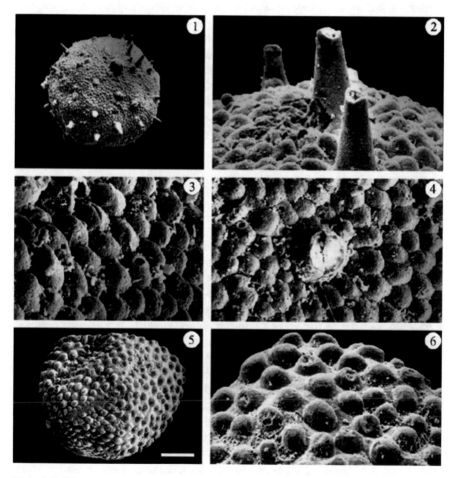

Fig. 9.5 SEM micrographs of three-dimensionally preserved Proterozoic acritarchs *Castaneasphaera* (*1–4*) and *Bacatisphaera* (*5, 6*) from China (reproduced after Chuanming et al. 2001, with permission ©John Wiley and Sons)

Ecology

The ecology/paleoecology of acritarchs is poorly known due to their uncertain biological affinity and rare occurrence in modern environments. Some of the well-preserved Proterozoic acritarchs are restricted to shallow water facies and associated with benthic, multicellular red and green algae, suggesting their benthic mode of life. The radially symmetrical, spinose morphology of the same well-preserved acritarchs, however, resembles many other planktic and heteroplanktic protistan cysts (Chuanming et al. 2001). Some broad distribution in relation to water masses is documented for different geological times. In the Cambrian, *Micrhystridium* occurred in inshore areas and *Skiagia* were found in offshore areas. The near-shore assemblage of Silurian comprises low diversity and low-to-moderate abundance of acritarchs, including *Leiosphaeridia*, *Veryhachium* and *Micrhystridium*. The offshore assemblage is of high diversity and moderate abundance. Low diversity assemblages of thick-walled sphaeromorphs characterize deep-water environments. The length of the processes of acritarchs increases in the offshore direction. The proximal and distal assemblages are also distinguishable by morphogroups. Simple, subspherical forms occur in proximal assemblages, thin-spined forms in intermediate settings, and thick-spined forms in distal settings (Jones 2006). The cold water, high latitudes in the Silurian are characterized by *Leiofusa*, *Neoveryhachium* and *Eupoikilofusa*, and the warmer, low latitudes contain *Deunffia*, *Domasia*, *Estiastra*, *Hogklintia* and *Pulvinosphaeridium* (Dorning 2005).

Geologic History

The first acritarchs are known to have existed in 1800 Ma Proterozoic, possessing vesicles of 20–200 µm in size. The peak of acritarch diversity was in the late Devonian and has dropped since the Carboniferous. The Precambrian acritarchs mostly belong to the group Sphaeromorphitae, which includes forms with spherical or subspherical vesicles, single layer walls and without processes or flanges. These acritarchs continue to the present day. The other group in the Precambrian was Acanthomorphitae. The acanthomorph acritarchs have spherical, ovoidal or ellipsoidal vesicles, are single- or double-walled, and possess processes. They continued to be an important element of the Paleozoic acritarchs. *Micrhystridium* appeared near the Ediacaran–Cambrian boundary, followed by rapid diversification of the acritarchs in the Early Cambrian. Several important taxa in the Cambrian to Devonian interval made this group valuable for biostratigraphy. The species *Polydryxium* and *Cymatiosphaera* were biostratigraphically significant in the Devonian period. There was a marked decline in acritarch taxa in the latest Devonian to Carboniferous. Small species of *Micrhystridium* and *Veryhachium* dominated the Triassic and continued to occur in the Mesozoic and Cenozoic. The acritarchs are not significant in the phytoplankton assemblages of the Mesozoic and Cenozoic eras.

9.3 Chitinozoa

Chitinozoans are an extinct group of organic-walled microfossils of uncertain affinity that ranged from the Ordovician to Devonian periods. Knowledge about their biology and ecology is limited and has been gathered indirectly from associated fauna and facies. They are interpreted to be pelagic and marine. They have been assigned to various groups, from protozoans to dinoflagellates and fungi, but the morphological similarities between the chitinozoans and metazoan eggs have led to the hypothesis that they were eggs of soft-bodied metazoans that had a planktic mode of life (Paris and Verniers 2005). The pyrolytic investigation of well-preserved Silurian chitinozoans did not yield any specific biomarker to reveal the biological affinity of the group (Dutta et al. 2007). Chitinozoa are particularly useful in biostratigraphy of deep marine sequences of the Ordovician to Devonian for which global biozones have been proposed. Chitinozoa are also good indicators of the thermal history of the sedimentary basins due to progressive darkening of the vesicle wall with an increase in temperature.

Morphology

The vesicles of chitinozoans are a flask-like structure formed of an organic wall (Fig. 9.6). The wall geochemistry indicates a predominance of aromatic compounds over the aliphatic compounds and absence of products diagnostic of chitin. It was,

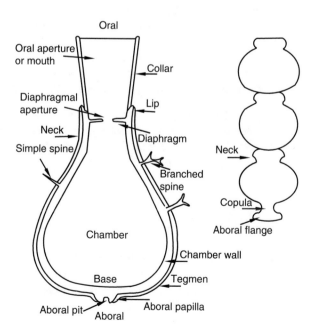

Fig. 9.6 General morphology of chitinozoa

therefore, unlikely that the biomacromolecules of chitinozoa were made of chitin prior to fossilization (Dutta et al. 2007). The chitinozoan vesicles are mostly 150–300 μm in length and may occasionally range up to 1500 μm. The wall is double layered and encloses an empty body chamber. The oral end bears an aperture that lies directly on the chamber or is produced in a neck. An operculum covers the aperture. The aboral end is broad and closed. The wall may be smooth, striate, tuberculate, spinose or hispid. The attributes of ornamentation, including shape, size, density and location, are of taxonomic importance. Some chitinozoans are joined in a long chain, with vesicles welded together at the operculum and the base. Various types of linkages are observed, including chain-like and planar.

Paleoecology

The environmental distribution of Chitinozoa is indirectly inferred from the associated taxa and sedimentary facies. Some significant observations about their geological distribution that give insight into their ecology are as follows (Paris and Verniers 2005):

1. They occur exclusively in marine rocks deposited in shallow water onto slope environments.
2. They are often present in good abundance in anoxic deposits of marine benthic fauna (e.g. in Silurian black shales).
3. The outer shelf and slope deposits yield abundant chitinozoans (several thousand specimens per gram of rock sample).
4. The geographical distribution of many species is much wider than that of the benthic organisms.
5. Some chitinozoans extend to all paleoclimatic belts.
6. Most of the chitinozoan genera and species have a worldwide distribution, comparable to the distribution of graptolites and conodonts.

Many Ordovician genera were cosmopolitan, for example, *Belonechitina*, *Conochitina* and *Cyathochitina*. Several of the genera had latitudinally restricted distribution. Species of *Eremochitina*, *Siphonochitina* and *Sagenachitina* were restricted to higher latitudes.

Geologic History

The organisms to which the chitinozoans are related evolved in the Tremadocian Stage of the Ordovician Period and lived until the latest part of the Famennian at the end of the Devonian Period. Chitinozoans were smooth walled and large in the

Early Ordovician. Among the first evolved taxa, the carinate forms expanded rapidly, but the siphonate forms soon declined and became extinct in the Middle Ordovician. Strongly ornamented forms appeared at the end of the Ordovician. The common genera of Ordovician, including *Siphonochitina*, *Sagenachitina*, *Herochitina* and *Acanthochitina*, became extinct at or near the Ordovician–Silurian boundary (Jansonius and Jenkins 1978). The Silurian was marked by species with spheroidal chambers and long necks. Chitinozoans declined in the Devonian and the assemblage in which they last appeared in the stratigraphic record comprised the species *Ramochitina*, *Cladochitina* and *Angochitina*.

9.4 Spores and Pollens

Spores and pollens are produced in the life cycles of different types of plant. Spores, produced by lower plants such as algae, fungi and ferns, appeared for the first time in the Silurian. Pollens are the male reproductive bodies of the seed plants. The first coniferous pollens described were from the Carboniferous and angiosperm pollens were from the early Cretaceous. The sensitivity of the plants to climate, topography, soil and groundwater, however, makes spores and pollens good indicators of paleoclimate. An important application of pollens is in Quaternary climate and archaeology. Due to the allochthonous nature of spores and pollens, their utility in marine sediments is limited. In terrestrial rocks, spores have been particularly important in age and correlation of coal seams. The change in colour of spore walls from pale-yellow through orange, brown and black is a useful measure of the thermal maturity of source rocks in hydrocarbon exploration. A brief description of the morphology of spores/pollens is given below, and the readers may refer to Traverse (2007) for their detailed morphology and distribution.

Morphology

Spores are tetrahedral, spheroidal or elongate, and occur both singly and in tetrads. A unit of four spores or pollens formed by one mother cell is a Tetrad. Spores or pollens arranged in a tetrahedron make a tetrahedral tetrad, and when arranged in a plane, a tetragonal tetrad (Fig. 9.7). The spores and pollens produced in tetrads may be differentiated into the proximal surface towards the centre of the tetrad and the distal surface away from the centre. The surface suture, called the laesurae, is the contact of the spores with their neighbours in the original tetrad from which they have been shed. Laesurae may be single, triradiate or may be absent, and, accordingly, the spores are monolete, trilete or alete (Fig. 9.7). The wall of spores and

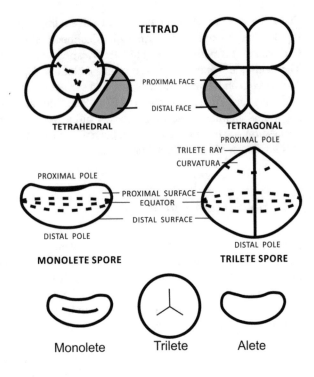

Fig. 9.7 General morphology and terminology of spores

pollens is made of an outer layer, termed the exine, and an inner layer, termed the intine. The exine in pollens is frequently divided into two layers, the endexine and the ektexine. The exine is composed of a highly resistant, complex compound of sporopollenin, while the intine is largely made of cellulose that is readily destroyed. A wide variety of shapes of pollen grains exists. Some of them have a wing-like structure called a saccus and the grains may be monosaccate or bisaccate. The number and type of apertures are the most distinguishing characteristics of pollen grains. The elongate or furrow-like apertures are called colpi and nearly circular ones are pores. Pollen grains may have no aperture (inaperturate), or colpate, porate or colporate (Fig. 9.8). Furthermore, there may be single or multiple pores and colpi. The surface of pollens may be smooth or characterized by pits, grooves and other elongate elements.

Geologic History

The plants producing simple trilete spores evolved in the Late Ordovician, from which the monolete spores were derived in the Silurian. The exine of trilete spores has shown diverse sculptures since the Late Silurian. The saccate pollens are

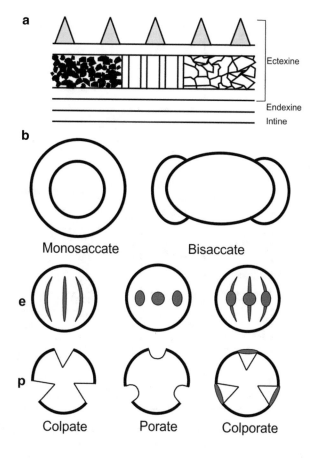

Fig. 9.8 Wall structure of spore/pollen (**a**) and general morphology of pollen grains. (**b**) Apertures are shown in equatorial (*e*) and polar (*p*) views

common among many gymnosperms. The monosaccates were more common than the bisaccates in the Carboniferous and Early Permian (Fig. 9.9). The complete array of morphologies exhibited by gymnosperm pollen had appeared by the mid-Mesozoic, but the monolete and trilete saccates and striate saccates of the Late Paleozoic and Early Mesozoic had disappeared. The Cretaceous gymnosperm pollens declined in the Cenozoic, subsequently attaining the modern aspect. In mid-latitudes and at the equator, gymnosperm pollens tend to be less diverse and abundant in the Cenozoic, including the present day when the angiosperm pollens dominate these assemblages. The angiosperm pollens appeared in the Early Cretaceous, although there are repeated reports of Jurassic and Triassic angiosperms. The earliest species, *Clavatipollenites hughesii*, was a monoaperturate and exine structure typical of the angiosperms (extexine differentiated into the foot

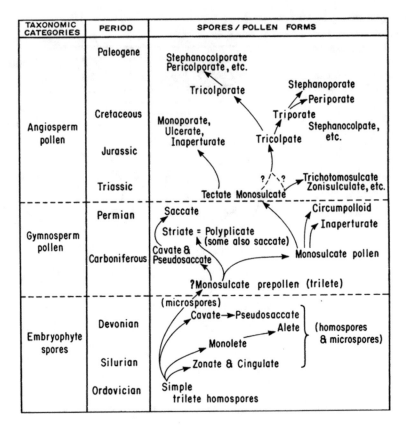

Fig. 9.9 Probable evolutionary pathways of spore/pollen morphological types (reproduced after Traverse 2007, with permission ©Springer Science+Business Media)

layer, columellar layer and a tectum). The angiosperm pollens are more useful in Cenozoic palynostratigraphy due to their proliferation in the Cenozoic. Some common pollens of the Eocene epoch are illustrated in Fig. 9.10. It should be remembered, however, that the ranges of spores and pollens are dependent on local climatic conditions. For example, the first appearance of tricolpate pollen is from the Neocomian to Cenomanian (Early Cretaceous), depending on the latitude, and the last appearance of Normapolles is in the Late Eocene to Early Oligocene (see Jansonius and McGregor 1996, Traverse 2007 for further reading on the evolution and distribution of spores and pollens).

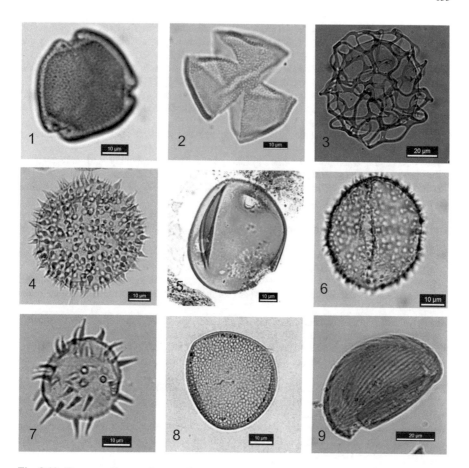

Fig. 9.10 Eocene pollens and spores from India: *Tribrevicolporites* (*1*) *Dipterocarpuspollenites* (*2*) *Retipollenites* (*3*) Spinizonocolpites (*4*) *Lakiapollis* (*5*) *Spinomonosulcites* (*6*) *Spinizonocolpites* (*7*) *Proxapertites* (*8*) *Schizaeoisporites* (*9*) (courtesy Jyoti Sharma)

References

Chuanming Z, Brasier MD, Yaosong X (2001) Three-dimensional phosphatic preservation of giant acritarchs from the terminal Proterozoic Doushantuo Formation in Guizhou and Hubei provinces, South China. Palaeontology 44:1157–1178

Crouch EM, Dickens GR, Brinkhuis H et al (2003) The *Apectodinium* acme and terrestrial discharge during the Paleocene–Eocene Thermal Maximum: new palynological, geochemical and calcareous nannoplankton observations at Tawanui, New Zealand. Palaeogeogr Palaeoclimatol Palaeoecol 194:387–403

De Vernal A, Rochon A, Turon J, Matthiessen J (1997) Organic-walled dinoflagellate cysts: Palynological tracers of sea-surface conditions in middle to high latitude marine environments. Geobios 30:905–920

Dorning KJ (2005) Acritarchs. In: Selley RC, Cocks LRM, Plimer IR (eds) Encyclopedia of geology, (5 Vols). Elsevier, Amsterdam, pp 418–428

Dutta S, Steiner M, Banerjee S et al (2006) *Chuaria circularis* from the early Mesoproterozoic Suket Shale, Vindhyan Supergroup, India: Insights from light and electron microscopy and pyrolysis-gas chromatography. J Earth Syst Sci 115:99–112

Dutta S, Brocke R, Hartkopf-Froder C et al (2007) Highly aromatic character of biogeomacromolecules in Chitinozoa: a spectroscopic and pyrolytic study. Org Geochem 38:1625–1642

Ellegaard M (2000) Variations in dinoflagellate cyst morphology under conditions of changing salinity during the last 2000 years in the Limfjord, Denmark. Rev Palaeobot Palynol 109:65–81

Evitt WR (1961) Observations on the morphology of fossil dinoflagellates. Micropaleontology 7:385–420

Evitt WR (1985) Sporopollenin dinoflagellate cysts: their morphology and interpretation. American Association of Stratigraphic Palynologists Foundation, Dallas

Jansonius J, Jenkins WA (1978) Chitinozoa. In: Haq BU, Boersma A (eds) Introduction to marine micropaleontology. Elsevier, Singapore, pp 341–357

Jones RW (2006) Applied palaeontology. Cambridge University Press, Cambridge

Paris F, Verniers J (2005) Chitinozoa. In: Selley RC, Cocks LRM, Plimer IR (eds) Encyclopedia of geology (5 vols.). Elsevier, Amsterdam, pp 428–440

Reichart G, Brinkhuis H (2003) Late Quaternary *Protoperidinium* cysts as indicators of paleoproductivity in the northern Arabian Sea. Mar Micropaleontol 49:303–315

Sarjeant WAS, Laealli T, Gaines G (1987) The cysts and skeletal elements of dinoflagellates: speculations on the ecological causes for their morphology and development. Micropaleontology 33:1–36

Sluijs A, Pross J, Brinkhuis H (2005) From greenhouse to icehouse; organic-walled dinoflagellate cysts as paleoenvironmental indicators in the Paleogene. Earth Sci Rev 68:281–315

Strother PK (2002) Acritarchs. In: Jansonius J, Mcgregor DC (eds) Palynology: principles and applications, 1st edn. American Association of Stratigraphic Palynologists Foundation, Texas, pp 81–106

Talyzina NM, Moldowan JM, Johannisson A, Fago FJ (2000) Affinities of Early Cambrian acritarchs studied by using microscopy, fluorescence flow cytometry and biomarkers. Rev Palaeobot Palynol 108:37–53

Williams DB (1971) The distribution of marine dinoflagellates in relation to physical and chemical conditions. In: Funnell BM, Riedel WR (eds) The micropalaeontology of oceans. Cambridge University Press, Cambridge, pp 91–95

Further Reading

Jansonius J, McGregor DC (eds) (1996) Palynology: principles and applications, vol 1–3. American Association of Stratigraphic Palynologists Foundation, Dallas

Sarjeant WAS (1974) Fossil and living dinoflagellates. Academic, London

Tappan H (1980) The paleobiology of plant protists. W H Freeman, San Francisco

Traverse A (2007) Paleopalynology, 2nd edn. Springer, The Netherlands

Part III
Applications

Chapter 10
Biostratigraphy

10.1 Introduction

Microfossils answer the most fundamental question of any geological investigation: "What is the age of the rock?" The basis of this answer is an understanding of the distribution of microfossils in time and space, the subject matter of *biostratigraphy*. The core of biostratigraphy is taxonomy. A biostratigrapher uses the short stratigraphic range of a fossil taxon or narrow stratigraphic intervals characterized by overlapping taxa to define a *biozone*. Biozones are used in dating and correlating the stratigraphic sequences. They may be interpolated with numerical age by radiometric dating, a process known as *biochronology*, and have an important bearing on the Geologic Time Scale.

Biostratigraphic zonations can be based on any kind of fossil fauna or flora. The relative values of different kinds of taxa, however, vary with geological age, depositional environment of the section under study and the kind of samples available for investigation. In subsurface stratigraphy, microfossils are invariably used for biozonation because of their small size, which is useful in the limited size of the samples recovered in drilling. One of the reasons why micropaleontology has reached its present level of specialization is because of the ability of microfossils to provide finer time slices of the stratigraphic column. The stimulus for this came from the oil industries, who initiated major research into foraminiferal biostratigraphy at the beginning of the 1940s. Another important phase in the development of microfossil-based biostratigraphy came in the 1960s through the initiative of oceanographic institutions engaged in the study of deep-sea sediments. Microfossils other than foraminifera also gained importance due to the Deep Sea Drilling Project (DSDP). The precision that the modern biochronology provides is due to the use of multiple groups of microfossils, including foraminifera, radiolaria, calcareous nanoplankton, silicoflagellates and diatoms. This chapter introduces some basic aspects of biostratigraphy and discusses the applications of microfossils in providing standard zonal schemes.

© Springer International Publishing Switzerland 2016
P.K. Saraswati, M.S. Srinivasan, *Micropaleontology*,
DOI 10.1007/978-3-319-14574-7_10

10.2   The Data for Biostratigraphy

One of the basic aims of biostratigraphy is to subdivide the fossiliferous sequences into finer units based on fossil contents. The finer units are too small for geological mapping purposes, but they are extremely useful for stratigraphic correlation. Two kinds of paleontological events are generally used in stratigraphy—the entry and the exit of a taxon in the rock record. The former is known as the First Appearance Datum (FAD) and the latter is called the Last Appearance Datum (LAD). Though the underlying principles in biostratigraphy are very simple, decision-making may often be quite complex due to the imperfection of the geological record. A number of uncertainties arise due to lack of a proper sampling plan, lack of confidence in taxonomic identification, influence of environmental change on the range of fossils and a differential rate of taxon evolution in different parts of the world. Simple observations like the presence or absence of a species in a sample should be questioned objectively, because they could arise due to various factors (Gradstein et al. 1985):

Taxon observed:	Is it its true presence?
	Is it due to geological reworking?
	Is there sampling contamination?
	Is there error in identification?

Taxon not observed:	Is taxon truly absent? If so,
	Is it its ecological exclusion?
	Is it its chronological exclusion?
	Taxon present but,
	Not recorded due to poor preservation?
	Is there an error in identification?
	Is it due to low population density?
	Is it due to high sedimentation rate?

The FAD and LAD represent the total range of existence of a taxon, from its evolution to extinction. The first occurrence (FO) and the last occurrence (LO) of a taxon in a locality, however, may be different from its FAD and LAD (Fig. 10.1). FO and LO define the biozone and FAD and LAD define the biochronozone. A major cause of error in stratigraphic interpretation is the assumption that FO and LO in a section are the same as FAD and LAD of the taxon. Due to hiatuses, poor preservation and imperfect sampling, the biozones may not be of the same age everywhere. The time elapsed from the evolution of a species in a province to its migration to another province may also cause the first occurrences to be of different times in

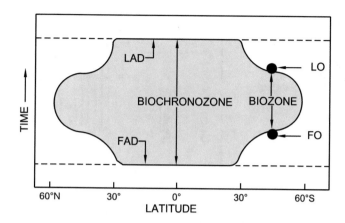

Fig. 10.1 The figure explains first occurrence (FO), last occurrence (LO), first appearance datum (FAD) and last appearance datum (LAD) of a species (redrawn after Olson and Thompson 2005, with permission © Springer Science + Business Media)

different regions. The first occurrences of the planktic foraminifer *Globorotalia truncatulinoides* in the Pacific and the Atlantic differ by 0.6 million years.

A species does not occur consistently from its FO to LO in a given stratigraphic section. Typically, it may show first occurrence, first consistent occurrence, peak occurrence, last consistent occurrence and last occurrence (Fig. 10.2). The sampling

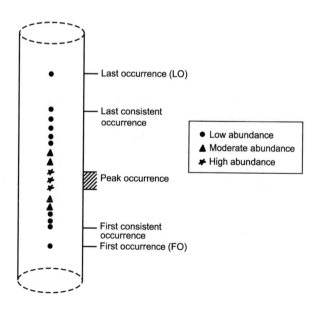

Fig. 10.2 A typical distribution of a species *x* in a core to illustrate different types of biostratigraphic data

Fig. 10.3 The unconformity, facies and condensation biases in the biostratigraphic ranges of species (redrawn after Holland 2000, with permission © Paleontological Society)

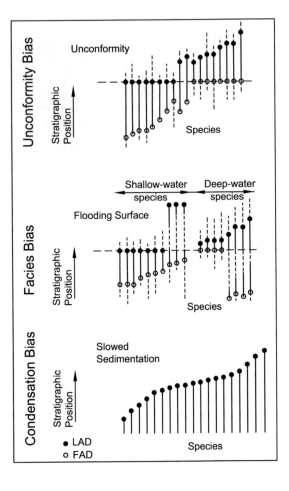

of such patchy stratigraphic distribution of taxa near origination or extinction produces artificially gradual origination and extinction patterns. Due to reduced sample size and rarity of species, the biostratigraphic ranges may show a gradual pattern of extinction even during periods of mass extinction, referred to as the *Signor–Lipps effect*. In addition to the sampling effect, unconformity, facies and slow sedimentation (condensation) bias the biostratigraphic range of species in an outcrop (Holland 2000). Unconformity causes FOs and LOs to cluster at unconformity due to non-deposition and erosion for a longer time. FOs and LOs may also cluster at surfaces of abrupt facies change and in zones of condensation where, because of slow sedimentation, the events appear more closely spaced (Fig. 10.3).

Fig. 10.4 Figure
explaining the Lazarus
effect. A foraminifer
species occurs in the lower
limestone bed deposited in
a marine environment,
disappears in the overlying
sandstone (non-marine)
and re-appears in the upper
limestone when the marine
environment returns

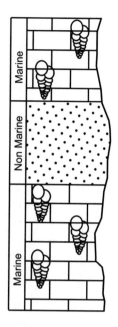

Sometimes, a taxon may disappear from a section not due to its extinction but due to its migration to another region because of ecological factors. It may reappear higher up in the section (Fig. 10.4) when the suitable condition returns (*Lazarus effect*). An objective evaluation of these factors in the first and last appearances of taxa is of critical importance in biostratigraphic interpretation.

The drill cutting samples are aggregates of certain intervals and are prone to contamination by borehole caving from the overlying strata. This affects the observed ranges of the fossil taxa (Fig. 10.5). Both the first and last occurrences of taxa can be determined with certainty in core and outcrop samples, but in drill cuttings, only the last occurrence can be correctly determined. Therefore, while working with drill cutting samples, only interval zones involving the LOs or the "first down-hole occurrences" are reliable.

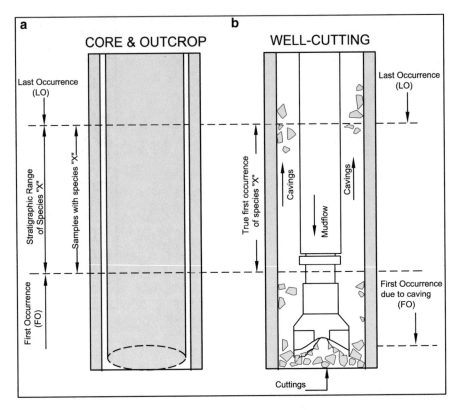

Fig. 10.5 The nature of sampling affects the observed first occurrence (FO) of a species. In core samples and outcrops (**a**), the first record at the bottom represents the true FO. In drill cuttings (**b**), the contamination by borehole caving from above may smear the FO of the species downward. The FO cannot be reliably determined in drill cuttings (redrawn after Olson and Thompson 2005, with permission © Springer Science + Business Media)

10.3   The Units of Biostratigraphy

A "zone" is a fundamental unit in biostratigraphy. It is a body of strata commonly characterized by the presence of certain fossil taxa. The most common types of zone are defined below and illustrated in Fig. 10.6.

Range Zone: defined by the stratigraphic range of a selected taxon in a fossil assemblage.

Concurrent Range Zone: the overlapping part of the ranges of two or more selected taxa.

Interval Zone: the stratigraphical interval between two successive biostratigraphic events.

Partial Range Zone: the partial range of a taxon between the last occurrence of a second and the first appearance of a third taxon.

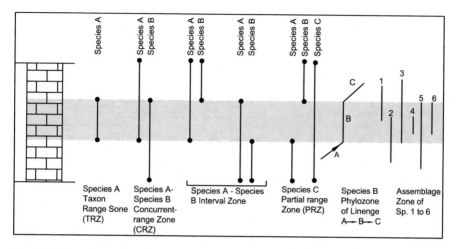

Fig. 10.6 The major types of biozone recognized in biostratigraphy

Assemblage Zone: defined by a distinctive assemblage of fossil taxa.
Phylo Zone: defined on the basis of the range of a taxon representing a segment of an evolutionary lineage (between the predecessor and the successor).
Acme Zone: the interval characterized by the maximum abundance of a particular species, unconcerned with the total range of the species.

Biostratigraphic boundaries are marked by bioevents, such as first appearance, last appearance, cryptogenic appearance and abrupt disappearance. When these bioevents are irreversible and isochronous over wide geographic regions, they are very useful and serve as biostratigraphic datum for correlation.

10.4   Microfossil-Based Biostratigraphic Zonations

Micropaleontologists started developing planktic foraminifera-based biostratigraphy for regional correlations in the 1940s. The works of HM. Bolli, FT. Banner, and WH. Blow in the late 1950s and early 1960s laid the foundation for planktic foraminiferal biostratigraphy. The DSDP was a turning point in the modern biostratigraphy and chronostratigraphy of the Cenozoic. The integration of different planktic microfossil zonation was a major achievement. Planktic foraminiferal biostratigraphy has undergone great refinement due to recovery of cores from wide-ranging latitudes and intensive study on evolutionary changes in planktic foraminifera. WH. Blow introduced a novel scheme to refer to the biozones by alphanumeric notation. P-zones were used for the Paleogene and N for the Neogene. A large number of biostratigraphers have invested time in refining and amending the zones established by their predecessors. It is a continuous process necessitated by the updated ranges of taxa and taxonomic revisions. The alphanumeric notation has

EOCENE TIME SCALE

TIME (Ma)	CHRONS	POLARITY	EPOCH	AGE	CALCAREOUS NANNOPLANKTON Martini (1971)	CALCAREOUS NANNOPLANKTON Bukry (1973-1975)	PLANKTONIC FORAMINIFERA Berggren et al. (1995)		PLANKTONIC FORAMINIFERA Berggren and Pearson (this work)	
31	C12n		OLIGOCENE	RUPELIAN	NP21	CP16 a	P19	T. ampliapertura IZ	O2	T. ampliapertura HOZ
32	C12r	EARLY					P18	Ch. cubensis - Pseudohastigerina spp. IZ	O1	P. naguewichiensis HOZ
33	C13n								E16	H. alabamensis HOZ
34	C13r	LATE	PRIABONIAN	NP19-20	CP15	P17 P16	L. T. cerroazulensis IZ / T. cunialensis / Cr. inflata CRZ	E15	G. index HOZ
35–36	C15r / C16n 2n			NP18		P15	Po. semiinvoluta IZ	E14	G. semiinvoluta HOZ
37–38	C17n 1 n	BARTONIAN	NP17		P14	Tr. rohri - M. spinulosa PRZ	E13	M. crassata HOZ
39–40	C18n 1 n 2n				P13	Gb. beckmanni TRZ	E12	O. beckmanni TRZ
41	C18r	EOCENE / MIDDLE	NP16	CP14 a	P12	M. lehneri PRZ	E11	M. lehneri PRZ
42	C19r						E10	A. topilensis PRZ
43	C20n	LUTETIAN	NP15 c / CP13 c		P11	Gb. kugleri / M. aragonensis CRZ	E9	G. kugleri / M. aragonensis CRZ
44–46	C20r		NP15 b / CP13 b					
47	C21n		NP15 a / CP13 a	P10	H. nuttalli IZ	E8	G. nuttalli LOZ
48	C21r		NP14 b / CP12 b				
49	C22n		NP14 a / CP12 a	P9	Pt. palmerae - H. nuttalli IZ	E7	A. cuneicamerata LOZ
50	C22r		NP13 / CP11	P8 M. aragonensis PRZ	E6 A. pentacamerata PRZ
51	C23n 2n	EARLY	YPRESIAN	NP12	CP10	P7	M. aragonensis / M. formosa CRZ	E5	M. aragonensis / M. subbotinae CRZ
52	C23r			NP11	CP9 b	P6 b	M. formosa / M. lensiformis / M. aragonensis ISZ	E4	M. formosa LOZ
53	C24n 3n			NP10 d c / CP9 a	P6 a	M. velascoensis - M. formosa / M. lensiformis ISZ	E3	M. marginodentata PRZ
54	C24r			NP10 b a			E1 E2	P. wilcoxensis / M. velascoensis CRZ / A. sibaiyaensis LOZ
55				NP9 b / CP8 b	P5	M. velascoensis IZ	P5	M. velascoensis PRZ
56	C25n	PALEOCENE LATE	THANETIAN	NP9 a / CP8 a	P4c	M. soldadoensis / Gl. pseudomenardii CRSZ		

Fig. 10.7 Planktic foraminiferal zones and calcareous nannoplankton zones of the Eocene, integrated with magnetic polarity chrons (reproduced after Berggren and Pearson 2005, with permission © Cushman Foundation for Foraminiferal Research)

also been revised and there are notations specific to each period, for example, P for Paleocene, E for Eocene, and so on (Fig. 10.7). The amendments are periodically updated, with Wade et al. (2011) having made the latest review. Standard zonal schemes based on calcareous nannofossils are also proposed. Two schemes of standard biozonation are proposed, coded as NP and NN (Nannoplankton Paleogene, Nannoplankton Neogene) in one scheme and CP and CN (Coccoliths Paleogene, Coccoliths Neogene) in the other. These biozonations have been very useful for dating and correlation. Other microplankton, including radiolaria, silicoflagellates, diatoms and dinoflagellates, have also gained importance in biostratigraphy. The planktic microfossil-based standard biozonation schemes have been integrated and tied with the magnetic polarity scale for the Cenozoic.

Planktic microfossil-based biostratigraphy works well with open marine and deep marine successions. It does not work optimally in shallow water carbonate sequences having a paucity of planktic microfossils but characterized by larger benthic foraminifera, and thus, shallow benthic zones are proposed (Serra-Kiel et al. 1998, Less et al. 2008, Scheibner and Speijer 2009). These zones are referred to as SBZ (Fig. 10.8). The interrelationship between different biostratigraphic zones and their integration with polarity chrons are given in the latest Geologic Time Scale, GTS-2012 (Gradstein et al. 2012).

	Stage		Zonation		Larger Benthic Foraminifera	
	Aquitanian		SBZ 24		*Miogypsina gr. gunteri/tani*	
OLIGOCENE	Chattian		SBZ 23		*Miogypsinoides, Lepidocyclinids, Nummulites bouillci*	
OLIGOCENE	Chattian		SBZ 22	b	*Lepidocyclinids, Nummulites vascus*	*Cycloclypeus*
OLIGOCENE	Rupelian		SBZ 22	a	*N. fichteli, N. bouillei*	*Bullalveolina*
OLIGOCENE	Rupelian		?			
OLIGOCENE	Rupelian		SBZ 21		*Nummulites vascus, N. fichteli*	
EOCENE	Priabonian		SBZ 20		*Nummulites retiatus, Heterostegina gracilis*	
EOCENE	Priabonian		SBZ 19		*Nummulites fabianii, N. garnieri, Discocyclina pratti minor* *Heterostegina reticulata mossanensis*	
EOCENE	Priabonian / Bartonian		SBZ 18		*Nummulites biedai, N.cyrenaicus, Heterostegina armenica armenica*	
EOCENE	Bartonian		SBZ 17		*Alveolina elongata* *Nummulites perforatus, N.brogniarti, N.biarritzensis*	
EOCENE	Lutetian		SBZ 16		*Nummulites herbi, N.aturicus, Assilina gigantea* *Discocyclina pulcra balatonica*	
EOCENE	Lutetian		SBZ 15		*Alveolina prorrecta, Nummulites millecaput*	
EOCENE	Lutetian		SBZ 14		*Alveolina munieri, Nummulites beneharnensis, N.boussacia* *Assilina spira spira*	
EOCENE	Lutetian		SBZ 13		*Aloveolina stipes, Nummulites laevigatus, N. uranensis*	
EOCENE	Ypresian	Cuisian	SBZ 12		*Alveolina violae, N. manfredi, N. campesinus* *Assilina major, A. cuvilleri*	
EOCENE	Ypresian	Cuisian	SBZ 11		*Alveolina cremae, Nummulites praelaevigatus, N. nitidus* *N. archiaci, Assilina laxispira*	
EOCENE	Ypresian	Cuisian	SBZ 10		*Alveolina schwageri, Nummulites burdigalensis burdigalensis* *N. planulatus, Assilina placentula, Discocyclina archiaci archiaci*	
EOCENE	Ypresian	Ilerdian	SBZ 9		*Alveolina trempina, Nummulites involutus, Assilina adrianensis*	
EOCENE	Ypresian	Ilerdian	SBZ 8		*Alveolina corbarica, Nummulites exilis, N. atacicus, Assilina leymeriei*	
EOCENE	Ypresian	Ilerdian	SBZ 7		*Alveolina moussoulensis, Nummulites praecursor*	
EOCENE	Ypresian	Ilerdian	SBZ 6		*Alveolina ellipsoidalis, Nummulites minervensis*	
EOCENE	Ypresian		SBZ 5		*Orbitolites gracilis, Alveolina vredenburgi, Nummulites gamardensis*	
PALEOCENE	Thanetian		SBZ 4		*Glomalveoiina levis, Nummulites catari, Assilina yvettae* *Hottingerina lucasi*	
PALEOCENE	Selandian		SBZ 3		*Glomalveolina primaeva*	
PALEOCENE	Selandian		SBZ 2		*Miscellanea globularis* *Lockhartia akbari*	
PALEOCENE	Danian		SBZ 1		*Bangiana hanseni, Laffitteina bibensis*	

Fig. 10.8 Palaeogene Shallow Benthic Zones and their characteristic larger foraminifera (after Serra-Kiel et al. 1998, Less et al. 2008, Scheibner and Speijer 2009, Gradstein et al. 2012)

10.5   Quantitative Biostratigraphy

A number of quantitative methods have been developed for biostratigraphic analysis. The quantitative methods become important when large data sets from hundreds of wells in a basin or thousands of wells across basins are examined to establish the optimum sequence of fossil events, such as first occurrence and last occurrence. Three of the sequencing methods have proved popular: graphic correlation, ranking and scaling (RASC) and constrained optimization (CONOP). Some basic aspects of these methods are discussed below. The Palaeontological Statistics (PAST) and StrataBugs are among the several software programs available commercially or in the public domain for carrying out quantitative biostratigraphic analyses.

Relative Biostratigraphic Value of Taxa

A stratigraphic section is likely to contain a large number of taxa. Their utility in biostratigraphy varies from species that yield little to no biostratigraphic information to the classical index fossil that is of maximum biostratigraphic importance. The basic biostratigraphic attributes of taxa are: vertical range, geographic distribution and amount of facies independence. In the conventional definition of an index species, ease of identification is another attribute. It is not included in quantifying their biostratigraphic value, because it is assumed that accurate and consistent taxonomy is fundamental to all biostratigraphy. Each of the above-mentioned attributes is standardized so that they are constrained to lie within a range of 0.0–1.0. The attributes listed above are calculated as follows:

$$\text{Vertical Range}\,(V) = \frac{\text{Thickness of sediment within section }m\text{ occupied by species }i}{\text{Total thickness of section }m}$$

$$\text{Facies Independence}\,(F) = \frac{\text{Number of facies in which species }i\text{ occurs}}{\text{Total number of facies present}}$$

$$\text{Geographic Persistence}\,(G) = \frac{\text{Number of sections or localities in which species }i\text{ occurs}}{\text{Total number of sections or localities}}$$

All of the biostratigraphic attributes range from 0.0 to 1.0. Classic index fossils have wide geographical distribution, and are both facies independent and short ranged. Such forms will have large values of G and F and a small value of V. Different workers have proposed the following three measures of RBV:

$$\text{RBV}_1 = F(1-V) + (1-F)G$$
$$\text{RBV}_2 = \left[F(1-V) + G(1-V)\right]/2$$
$$\text{RBV}_3 = (1-V)$$

The values of the RBVs vary from 0 to 1. An RBV of near 0 indicates a fossil with no useful biostratigraphic information, while an RBV of almost 1 indicates a taxon which is of highest value for biostratigraphic correlation. The measure RBV_3 is applied to data sets of DSDP logs where many of the taxa are geographically widespread and facies independent. In quantitative correlation, RBVs provide weights so that the presence of species can be weighted in finding the amount of similarity in samples. Multivariate analysis of the weighted similarity matrices can be used to find the assemblage zones (see Gradstein et al. 1985 for more details).

Graphic Correlation

This is a graphical and semi-quantitative approach to stratigraphic correlation of events. Shaw (1964) introduced this method based on the simple concept that,

Thickness of rock = Rate of rock accumulation × Total time of accumulation

The procedure involves the compilation of first and last occurrences of all taxa in several sections. The best section is chosen as the reference section, being the one that is well sampled and has many species present at many stratigraphic levels. Each of the two sections to be correlated is first plotted on a different axis of a graph. The units of the graph axes represent distances from the base of the section to its top. The occurrences of each species in the two sections are represented on the graph (scatter plot) by points representing the top of its range, the base of its range or both. A regression line known as the line of correlation is fitted onto the scatter plot (Fig. 10.9). If the species ranges in the two sections are identical, the line of correlation will form an

Fig. 10.9 Graphic correlation plot of datum levels in DSDP sites 587 and 588 (Southwest Pacific). The change in the slope of the line of correlation indicates a change in the rate of sedimentation (redrawn after Srinivasan and Sinha 1991, with permission © Geological society of India)

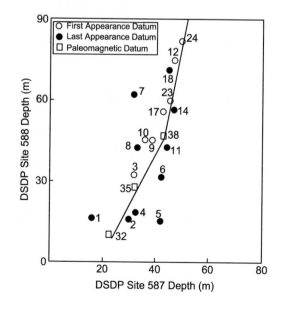

angle of 45° with each axis. The line of correlation is not necessarily a single straight line; it can be a series of interconnected line segments. Several situations are possible in correlation. A composite standard section (CSS) is prepared based on several local sections. To construct a CSS, the most complete section (the reference section) is correlated to each local section with the reference section plotted as the X-axis and the local section on the Y-axis. All event data of the local section are projected onto the line of correlation and then down to the reference section. The positions of the earliest first occurrences and latest last occurrences for each species are preserved. After correlating several sections with the reference section (which is usually done by computer as an iterative process), it achieves the status of a composite standard section. The graphic correlation is very useful in biostratigraphy, as its enables us to locate the seat of evolution and the routes of migration of fossil biota. It reveals the diachroneity of biotic appearances and disappearances.

Ranking and Scaling (RASC)

The RASC is a probabilistic method that uses fossil event order in several locations (or wells) to construct a most likely sequence of events. In simple terms, consider the first occurrence of sp X and sp Y and compare the order of the two events at all localities. The order that occurs most commonly is the most probable sequence of events. The method also uses the scores of cross-over (incorrect order) from well to well for all pairs of events. After ranking, the events are arranged from oldest to youngest. Once events are correctly ordered, stratigraphic distances are estimated between the events. RASC finds the average stratigraphic position of the first and last occurrence events. Both RASC and the below-discussed CONOP method handle large data sets and the analysis is automated. A data set of more than 50 taxa and over ten wells is required for these methods to function optimally (see Agterberg and Gradstein 1999 for detailed discussion of the method).

Constrained Optimization (CONOP)

This is an improvement of graphic correlation and can handle large data sets of multiple wells. This method attempts to find the maximum or most common stratigraphic ranges of taxa. It quantifies the misfit in the order of first and last appearances of taxa from one section to the other and the composite section, and then uses an optimizing algorithm for the global composite sequence of events. CONOP can model either average ranges or total ranges (Gradstein et al. 2008).

References

Agterberg FP, Gradstein FM (1999) The RASC method for ranking and scaling of biostratigraphic events. Earth Sci Rev 46:1–25

Berggren WA, Pearson PN (2005) A revised tropical to subtropical Paleogene planktonic foraminiferal zonation. J Foraminifer Res 35:279–298

Gradstein FM, Agterberg FP, Brower JC, Schwarzacher WS (1985) Quantitative stratigraphy. D Reidel, Dordrecht

Gradstein FM, Bowman A, Lugowski A, Hammer O (2008) Increasing resolution in exploration biostratigraphy – Part I. In: Bonham-Carter GF, Cheng Q (eds) Progress in geomathematics. Springer, Berlin

Gradstein FM, Ogg JG, Schmitz MD, Ogg GM (2012) The geologic time scale 2012, 2 vols. Elsevier, Amsterdam

Holland SM (2000) The quality of the fossil record: a sequence stratigraphic perspective. Paleobiology 26:148–168

Less G, Ozcan E, Papazzoni CA, Stockar R (2008) The Middle to late Eocene evolution of nummulitid foraminifer *Heterostegina* in the western Tethys. Acta Palaeontol Pol 53(2):317–350

Olson HC, Thompson PR (2005) Sequence biostratigraphy with examples from the Plio-Pleistocene and Quaternary. In: Koutsoukos EAM (ed) Applied stratigraphy. Springer, The Netherlands, pp 227–247

Scheibner C, Speijer RP (2009) Recalibration of the Tethyan shallow-benthic zonation across the Paleocene–Eocene boundary: the Egyptian record. Geol Acta 7:195–214

Serra-Kiel J, Hottinger L, Caus E, Drobne K, Ferrandez C, Jauhri AK, Less G, Pavlovec R, Pignatti J, Samso JM, Schaub H, Sirel E, Strougo A, Tambareau Y, Tosquella J, Zakrevskaya E (1998) Larger foraminiferal biostratigraphy of the Tethyan Paleocene and Eocene. Bull Soc Geol Fr 169:281–299

Shaw AB (1964) Time in stratigraphy. McGraw Hill, New York

Srinivasan MS, Sinha DK (1991) Improved correlation of the Late Neogene planktonic foraminiferal datum planes in the equatorial to cool subtropical DSDP sites, Southwest Pacific: Application of the graphic correlation method. Geol Soc India Memoir 20:55–93

Wade PS, Pearson PN, Berggren WA, Palike H (2011) Review and revision of Cenozoic tropical planktonic foraminiferal biostratigraphy and calibration to the geomagnetic polarity and astronomical time scale. Earth Sci Rev 104:111–142

Further Reading

Gradstein FM, Ogg JG, Schmitz MD, Ogg GM (2012) The geologic time scale 2012, (2 vols.). Elsevier, Amsterdam

Koutsoukos EAM (ed) (2005) Applied stratigraphy. Springer, The Netherlands

McGowran B (2005) Biostratigraphy – microfossils and geological time. Cambridge University Press, Cambridge

Chapter 11
Paleoenvironment and Paleoclimate

11.1 Introduction

Microfossils occur in sediments of all ages and under practically all environmental conditions. Their abundance in a sample in comparison to mega-fossils makes them a true representative of the biological community of an environment. A host of physical, chemical and biological factors of the environment control the distribution of microfauna. As a result, there are characteristic assemblages of microfauna corresponding to specific environments of deposition. Furthermore, the oxygen and carbon isotopes and trace elements in shells of several microfossils have enabled quantitative estimation of environmental parameters. The key points in the micropaleontologic approach to interpretation of paleoenvironment and paleoclimate are as follows:

1. The basis of interpretation is environmental controls on the present-day distribution of microfauna (uniformitarian approach).
2. The more specific and limited the criteria derived from the present-day fauna, the lower the probability of finding analogous criteria in ancient assemblages.
3. The older the faunal assemblage, the less likely that its environmental distribution will be similar to the present-day fauna.
4. The environmental conditions limiting the distribution of microfauna vary. The environmental tolerance may be large for a group, but the specific requirements of species within the group may be very limited. For example, the ostracoda are

© Springer International Publishing Switzerland 2016
P.K. Saraswati, M.S. Srinivasan, *Micropaleontology*,
DOI 10.1007/978-3-319-14574-7_11

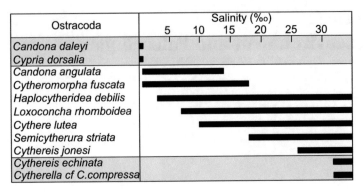

Fig. 11.1 Narrow to wide tolerance for salinity by different species of ostracoda (data after Bignot 1985; Keen 1993)

euryhaline, living in fresh to hypersaline conditions, but the salinity tolerance of some species is very limited (Fig. 11.1).

5. The pre-requisite for paleoenvironmental interpretation is a carefully carried out taphonomic evaluation of fossil assemblages. Preferential preservation in certain settings, such as marshes and deep sea, may have a marked effect on the original assemblages.

The diagnostic microfossils of different environments are listed in Table 11.1. Except for the spores and pollens, which are terrestrial, the remaining microfossils are aquatic and mainly marine. The general ecology of these microfossils has been discussed previously. This chapter discusses the major aspects of paleoenvironmental interpretation, including the estimation of water depth (paleobathymetry), sea-level change, oceanic productivity, oxygen content and paleoclimate.

Table 11.1 Diagnostic microfossils of the continental, marginal marine and marine environments

Environment	Diagnostic microfossils
Continental	Spores and pollens, diatoms, ostracoda
Marginal marine	Foraminifera, ostracoda, dinoflagellates, algae
Marine	Foraminifera, ostracoda, pteropods, diatoms, dinoflagellates, bryozoa, acritarch, radiolarians, calcareous nannoplankton

11.2   Paleobathymetry and Sea-Level Change

The first-order discrimination of marine environments is achieved by the ternary plot of Murray (2006). This is based on wall types in foraminifera. The three major types of wall structure in foraminifera are agglutinated, calcareous hyaline and calcareous porcelaneous. The method involves plotting the percentage abundance of the three wall types in the sample on the ternary plot. The marginal marine, shelf sea and deep-sea samples can be distinctly distinguished, despite some overlaps (Fig. 11.2). The samples from marshes occupy the whole triangle. The Fischer's α-diversity (Sect. 5.3), taking the count of foraminiferal species and number of individuals, is also informative and complements the interpretation based on the ternary plot. The cross-plot of α-diversity and information function H further distinguishes marginal marine, shelf sea and deep-sea environments, although some overlap exists (Fig. 11.3).

The estimation of water depth requires a detailed identification of foraminiferal assemblages. A number of environmental variables, such as light, organic flux and oxygen, change with depth. A general variation in a foraminiferal assemblage with depth may be due to the influence of one or more of these variables. It may, however, be noted that the foraminiferal genera are not restricted to a depth zone but occur persistently in the zone, and their abundance changes progressively from one depth zone to the next. Culver (1988) used a comprehensive database of foraminifera in the Gulf of Mexico to recognize 14 depth zones that are the basis of paleobathymetric interpretation. A simplified depth distribution of the foraminiferal genera is

Fig. 11.2 Ternary plot of foraminifera for discrimination of marsh (*whole field*), marginal marine (*grey*), shelf seas (*hatched*) and deep-sea (*dotted*) environments. Only the main fields for shelf seas and deep seas are shown (from Murray 2006, with permission © Cambridge University Press)

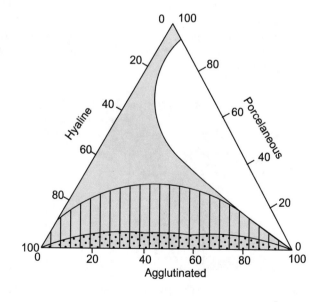

Fig. 11.3 Bivariate plot of
Fisher-alpha and H-index
for foraminifera from
different environments
(redrawn after Murray
2006, with permission ©
Cambridge University
Press)

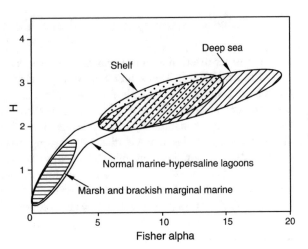

shown in Fig. 11.4, and characteristic assemblages of neritic (continental shelf),
bathyal (continental slope) and abyssal biofacies are given below:

Marginal Marine (including estuaries and lagoon): The agglutinated taxa
   Jadammina, *Miliammina* and *Trochammina* are diagnostic of high marsh.
   Ammobaculites, *Ammotium* and *Miliammina* characterize estuaries and brackish
   lagoons. Calcareous foraminifera *Ammonia* and *Elphidium* occur in lower
   reaches of estuaries.

Inner to Middle Neritic (up to 100 m): This is characterized by *Ammonia*, *Bigenerina*,
   Bolivina, *Cibicides*, *Discorbis*, *Elphidium*, *Nonionella*, *Quinqueloculina* and
   Textularia. Planktic foraminifera constitute <10 % of the assemblage.

Outer Neritic (100–200 m): The common benthic genera include *Bolivina*, *Bulimina*,
   Cancris, *Cassidulina*, *Lenticulina*, *Marginulina*, *Reusella*, *Sigmoilina* and
   Uvigerina. The planktic foraminifera may constitute about 50 % of the total
   population.

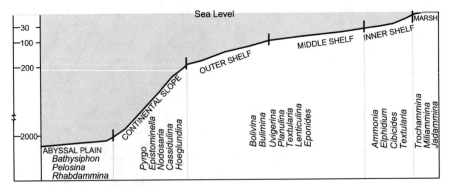

Fig. 11.4 Common assemblages of benthic foraminifera in marginal marine, continental shelf and
continental slope environments

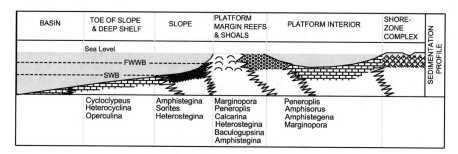

BASIN	TOE OF SLOPE & DEEP SHELF	SLOPE	PLATFORM MARGIN REEFS & SHOALS	PLATFORM INTERIOR	SHORE-ZONE COMPLEX	
	Cycloclypeus Heterocyclina Operculina	Amphistegina Sorites Heterostegina	Marginopora Peneroplis Calcarina Heterostegina Baculogupsina Amphistegina	Peneroplis Amphisorus Amphistegena Marginopora		

Fig. 11.5 Distribution of larger benthic foraminifera in modern reef and associated environments (data after Hallock and Glenn 1986)

Bathyal (>200 m): This assemblage commonly comprises *Bolivina, Bulimina, Cassidulina, Chilostomella, Epistominella, Glomospira, Gyroidinoides, Haplophragmoides, Hoeglundina, Nodosaria, Pullenia, Pyrgo, Uvigerina* and *Valvulineria.* The planktic foraminifera may be 80 % or more of the population.

Abyssal plain: Simple agglutinated foraminifera characterize the abyssal plain. The deepest water genera, recorded from the Mariana trench (~11,000 m), include *Hormosina, Lagenamminadi, Reophax* and *Rhabdammina.*

Larger benthic foraminifera are also depth controlled. *Peneroplis, Amphisorus, Marginopora, Baculogypsina* and *Calcarina* characterize the reef-flat and back-reef lagoons in the present-day coral reefs (Fig. 11.5). The deeper water fore-reef regions typically contain *Heterostegina, Operculina* and *Cycloclypeus.* Some of the taxa may have wider depth distribution, but they may show morphotypic variation with depth. *Amphistegina* shows progressively inflated shells with increasing energy and intensity of light (Fig. 11.6). In the Eocene carbonate ramps, *Alveolina* and *Orbitolites* occurred in the inner ramps and *Nummulites, Assilina* and *Discocyclina* occurred with the progressive deepening of the ramps. The assemblages *Sorites–Peneroplis–Marginopora, Borelis–Alveolinella–Austrotrillina, Lepidocyclina–*

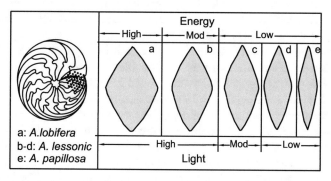

Fig. 11.6 Change in shape of *Amphistegina* with light and energy of the environment (data after Hallock and Glenn 1986)

Heterostegina and *Cycloclypeus* respectively characterized inner to outer ramps in the Oligo-Miocene (Beavington-Penney and Racey 2004).

The ratio of planktic and benthic foraminifera (P/B) is another criterion for estimating depth. The planktic foraminifera live in open ocean environments, usually away from the coast. The proportion of planktic to benthic foraminifera (in %) varies from the inner shelf to the continental slope as follows:

Inner shelf	<20:>80
Middle shelf	10–60:90–40
Outer shelf	40–70 : 60–30
Upper cont. slope	>70:<30

Van der Zwaan et al. (1990) argued that organic matter, varying with depth in the ocean, is of ecological importance for both planktic and epifaunal benthic foraminifera, but not for the infaunal benthic foraminifera. They re-examined the P/B ratio by removing the infaunal taxa and found an improved relationship with depth as follows:

$$D = e^{\left(3.58718 + \left(0.03534\, \%P^*\right)\right)},$$

where D is depth (in metre) and $\%P^*$ is the redefined P/B ratio (by excluding the infaunal species). The equation is useful in estimation of depths between 30 and 1250 m.

The morphotypes of benthic foraminifera show a distinct depth distribution (Jones and Charnock 1985, Corliss and Fois 1990). The planoconvex type, comprising the species *Cibicides*, *Discorbis* and *Planulina*, constitutes <10 % at depths shallower than 1000 m, but increases up to 60 % at depths >1000 m. The flat, ovoid, tapered and cylindrical morphotypes (mainly containing the species *Bulimina*, *Bolivina*, *Trifarina*, *Reusella*, *Nodosoria*, *Virgulina* and *Cassidulina*) have maximum values in the upper 2000 m and low values in the 2000–4000 m water depth. The infaunal taxa dominate in shallow waters from 100 to 1300 m and the epifaunal taxa dominate the deeper waters of >2000 m. The depth distribution of agglutinated foraminifera morphogroups is also of potential significance in paleobathymetric interpretation. In general, multilocular, planispiral/trochospiral and lenticular forms are common in shelf and marginal marine environments while unilocular, tubular or branching forms are characteristic of the deep sea.

For the reconstruction of sea-level changes, multiple approaches are employed. The sample-wise distribution of generic assemblages, abundances of the recorded genera, species diversity and P/B ratios plotted in stratigraphic order provide an estimate of paleo-water depths. Unlike other quantitative analysis, it is difficult to judge an exact error in estimation of depth. The estimation of depth is more precise for the inner to middle shelf and the error increases in the outer shelf and bathyal environments. In the eustatic sea-level curve for the past 100 Ma, Miller et al. (2005) suggested an error of ±15 m for the inner shelf, ±30 m for the middle shelf, ±50 m for the outer shelf and ±200 m for the slope environments. The $\delta^{18}O$ of foraminiferal shells complements the sea-level curve constructed based on assemblages

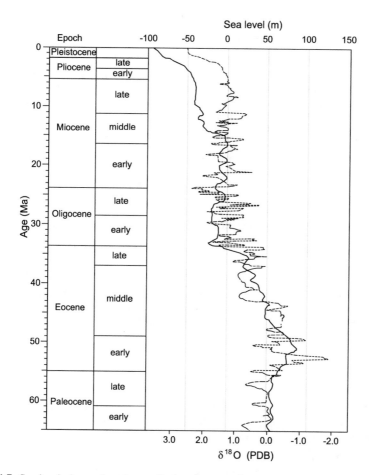

Fig. 11.7 Sea-level change (*continuous line*) and oxygen isotope (*dashed-line*) curve for the Cenozoic Era (redrawn after Miller et al. 2011, with permission © The Oceanographic Society)

of foraminifera. This is because the ice volume effect in $\delta^{18}O$ is due to a change in the volume of ice that also causes a change in the sea level. Notwithstanding the limitations and uncertainty in oxygen isotope methodology, there is a broad similarity in the sea-level curve and the $\delta^{18}O$ curve (Fig. 11.7).

11.3   Organic Flux and Dissolved Oxygen

The abundance of foraminifera in the deep sea is closely linked to levels of organic matter input and dissolved oxygen concentration, although separating the effects of the two factors is not easy (Gooday 1994). The TROX model postulates that the

Table 11.2 Major deep-sea benthic foraminiferal morphotypes and their distribution in relation to oxygen content and organic matter (after Gooday 1994)

Morphotypes	Other characteristics	Environment/ microhabitat	Example
Planoconvex trochospiral, Rounded trochospiral, Biconvex trochospiral, Milioline	Large pores absent or on one side; low surface/volume ratio	Well oxygenated Epifaunal	*Cibicides, Cibicidoides, Gyroidina, Oridorsalis, Pyrgo*
Planoconvex trochospiral, Biconvex trochospiral, Spherical, Tapered cylindrical	Small tests, thin hyaline walls	Phytodetritus in open ocean Epifaunal	*Epistominella, Alabaminella, Fursenkoina, Globoccassidulina*
Tapered cylindrical, Tapered flattened, Rounded planispiral, Cylindrical planispiral	Pores evenly distributed	Low oxygen and/or high organic matter Infaunal	*Uvigerina, Trifarina, Brizalina, Bolivina, Pullenia, Nonionella, Globobulimina, Fursenkoina*
Tapered cylindrical, Rounded planispiral, Cylindrical planispiral		Oxygenated and high organic matter Infaunal	*Chilostomella, Globobulimina, Melonis, Uvigerina*

combination of the two factors limits the abundance of infaunal foraminifera (see Box 5.1, Chap. 5). Organic flux and oxygen concentration also control the microhabitat of species, as reflected in the morphotypes (Table 11.2). The overall shape of the tests, pores, thickness of the test wall and microhabitats are, thus, highly adaptive. The well-oxygenated epifaunal microhabitats in the deep sea are characterized by planoconvex to biconvex tests, low surface area to volume ratios, and large pores that are either absent or confined to one side. The typical taxa having these morphological attributes include *Cibicidoides*, *Gyroidina*, *Oridorsalis* and *Pyrgo*. Sen Gupta and Machain-Castillo (1993) remarked that there is no modern species of benthic foraminifera exclusive to low-oxygen bottom waters or pore waters. However, they found that the bathyal oxygen minimum zones (OMZ; oxygen concentration of 0–2.0 ml/l) in the Indian and Pacific Oceans are characterized by low diversity and high dominance of small, thin-shelled forms of *Bolivina*, *Buliminella*, *Globobulimina*, *Uvigerina*, *Epistominella* and *Cassidulina*. Large pores and more numerous pores are characteristic of foraminifera from the low-oxygen zone.

Kaiho (1994) observed that oxygen concentration plays a major role in controlling foraminiferal assemblage and their morphologic characteristics. He proposed a *dissolved oxygen index* based on benthic foraminifera and found a good correlation

Table 11.3 Indicator benthic foraminifera of oxic, suboxic and dysoxic environments (compiled from Kaiho 1994)

Environments	Oxygen concentration (ml/l O_2)	Benthic foraminifera	Morphological characteristics
Oxic	>1.5	Species of *Cibicidoides, Nuttallides, Stensioina, Gavelinella, Globocassidulina*	Thick walls, large tests, planoconvex, biconvex, rounded trochospiral, spherical, epifaunal under high oxygen bottom-water but absent in low-oxygen environment
Suboxic	0.3–1.5	Small specimens of oxic species, *Lenticulina, Nodosaria, Dentalina, Pleurostomella, Bulimina, Stilostomella, Uvigerina, Oridorsalis, Gyroidina, Hoeglundina*	Small and/or thin walled, rounded planispiral, flat ovoid, spherical, biconvex trochospiral
Dysoxic	0.1–0.3	Species of *Bolivina, Chilostomella, Fursenkoina, Globobulimina*	Thin wall, small, flat or elongate tapered, high porosity tests

between the proposed index and the concentration of oxygen in the overlying water. The index is calculated as,

$$\left[O/(O+D)\right]\times 100 \text{ for } O > 0 \text{ and,}$$

$$\left\{\left[I/(I+D)\right]-1\right\}\times 50 \text{ for } O = 0 \text{ and } D+I > 0,$$

where O, D and I are numbers of specimens of oxic, dysoxic and suboxic indicators, respectively. Some of the indicator species are listed in Table 11.3, but the reader should refer to Kaiho (op. cit.) for a complete list of species to calculate the index. The oxygen index provides a quantitative estimate of the level of oxygenation in the environment (Table 11.4).

Table 11.4 Foraminifera-based oxygen index and the corresponding level of oxygen concentration (after Kaiho 1994)

Oxygen condition	Oxygen level (ml/l)	Oxygen index
High oxic	3.0–6.0+	50–100
Low oxic	1.5–3.0	0–50
Suboxic	0.3–1.5	−40 to 0
Dysoxic	0.1–0.3	−50 to −40
Anoxic	0.0–0.1	−55

11.4   Ocean Eutrophication

The term eutrophic refers to high levels of nutrient availability and the associated increase in primary productivity in the ocean. The nutrients for the marine micro-organisms are principally the dissolved phosphate and nitrate in the seawater. Eutrophication is defined as an increase in the rate of supply of organic matter to an ecosystem; the level of organic carbon supply in a eutrophic condition is 301–500 g C/m^2 years (Nixon 1995). Eutrophication, a consequent change in productivity and a lowering of the levels of atmospheric CO_2 are related issues in paleoclimate reconstruction. Increased fluvial runoff and upwelling seawater are among the various factors promoting ocean eutrophication. During upwelling, there is a vertical movement of cold and nutrient-enriched bottom water to surface water. The study of responses of modern planktic foraminifera, diatoms, radiolarians and dinoflagellates to upwelling water has led to the development of indicator species of upwelling and eutrophication. There are several groups of microfossils, as described below, that provide evidence of past ocean eutrophication (see Brasier 1995 and Murray 1995 for a detailed review).

Planktic foraminifera: A subpolar planktic foraminifer *Globigerina bulloides* is a well-known indicator of upwelling and eutrophy in modern seas. In south Asia, upwelling is closely linked to strong monsoon winds and, therefore, a census count of this species has become a routine procedure in paleomonsoon reconstruction. The other markers of upwelling include *Globigerinoides ruber*, *Globigerinoides sacculifer*, *Neogloboquadrina dutertrei* and *Globigerinita glutinata*.

Diatoms and Radiolarians: The fluxes of diatoms and radiolarians are indicators of high productivity in the modern sea. The diatoms *Chaetoceros* and *Thalassiosira* are markedly present in upwelling regions. The diatoms increase in number in upwelling areas and radiolarians increase in the surrounding regions. The diatom/radiolarian ratio, therefore, maps the core of the upwelling zone.

Dinoflagellates: The organic-walled plankton dinoflagellates also increase during eutrophication. The present-day upwelling areas have a higher abundance of cooler water peridiniacean (P) cysts compared to the gonyaulacean (G) cysts dominating in areas of weak upwelling. Variation in the P/G ratio, therefore, marks changes in the intensity of upwelling.

11.5   Ocean Acidification

The uptake of anthropogenic CO_2 by an ocean decreases the pH of the seawater and lowers the saturation state of calcite and aragonite. The process, termed ocean acidification, is expected to have detrimental consequences for marine organisms, including corals, foraminifera and other $CaCO_3$-secreting organisms (Zeebe et al. 2008). The critical value of pH below which the calcifying organisms will find it difficult to calcify or survive in the ocean water is not currently known. But it is well recognized that rising pCO_2 and the consequent ocean acidification will severely impair many ecosystems, most notably the coral reefs. The effect of ocean acidification on marine biota is a nascent field of study. The ways in which incremental change in pH will affect the calcium carbonate secreting organisms is being investigated in nature and laboratory cultures. The culture experiments have shown that net calcification in the species of larger benthic foraminifera tends to decrease at elevated pCO_2 levels of 970 µatm in comparison to the present level of 360 µatm (Fujita et al. 2011). During the Paleocene–Eocene hyperthermal events, when atmospheric CO_2 reached above 1000 ppm (Pearson and Palmer 2000), there was massive sea floor carbonate dissolution and the CCD shoaled rapidly by more than 2 km (Zachos et al. 2005). The benthic extinction accompanying abrupt warming at the P–E boundary is attributed to carbonate under-saturation in the deep sea, impeding calcification by marine organisms. There are several other indications of ocean acidification in the geologic past, the latest being the deglaciation in the Late Pleistocene, marked by a 30 % rise in CO_2 between 17.8 and 11.6k years BP. It is estimated that the CO_2 increased from 189 to 265 µatm, and the boron isotopes of planktic foraminifera suggest a decrease in sea surface pH by 0.15 ± 0.05 of a unit (Honisch et al. 2012).

11.6 Environmental Monitoring

There is a growing concern about the impact of anthropogenic activities on ecosystems. It is necessary that the question of environmental and/or climate change and their impact on biota be addressed with sound scientific backing. Due to an understanding of the long-term effect of environmental change on biotic communities, paleontologists have better insight into the complexity of natural versus human-induced environmental changes and the varying response of biota to such changes. Several studies have demonstrated the usefulness of foraminifera, ostracoda, dinoflagellates, diatoms and other microfauna as indicators of heavy metal pollution, sewage discharge, quality of lake water and eutrophication due to industrial pollution. Martin (2000) explains the applications of micropaleontology in environmental geology with several case studies.

Anthropogenic pollutions change the composition, abundance and diversity of microfaunal assemblages and deform the shells. The coastal waters of the industrial city of Mumbai in India are affected by heavy metal contamination from affluent wastes. The different genera of foraminifera respond differently to metal contamination, *Lagena* being least affected while *Ammonia* and *Cibicides* are affected most by the level of contamination (Banerji 1990). Although various degrees of morphological abnormalities in the foraminiferal specimens were documented from the region, no definitive relationship was observed with the concentration of heavy metals. The morphological deformities in the foraminiferal tests include aberrant chambers, twisted chamber arrangement, protuberances, twinning and poor development of last whorls, among several such features. The deformations, however, are not necessarily due to heavy metal contamination. Several species are recorded as having high proportions of deformed tests due to low salinity (Yanko et al. 1998). Oil spills also reduce the overall abundance of foraminifera. The infaunal and opportunistic species increase near such stressed environments (Mojtahid et al. 2006). The diatoms and dinoflagellates have also proved to be good bio-indicators of eutrophication and, thus, help monitor the quality of lake water and coastal environments.

An important application of micropaleontology is suggested in monitoring the health of coral reefs. Hallock et al. (2003) proposed a foraminiferal index (FI) to indicate the quality of water suitable for healthy growth of corals. It is defined as,

$$FI = (10 \times Ps) + (Po) + (2 \times Ph),$$

where Ps is the proportion of symbiont-bearing larger benthic foraminifera, Po is the proportion of opportunistic foraminifera and Ph is the proportion of the other foraminifera. The FI value is interpreted as follows:

$FI>4$	Environment conducive to reef growth
$2<FI<4$	Environment marginal for reef growth and suitable for recovery
$FI<2$	Stressed environment, unsuitable for reef growth

11.7   Paleoclimate

Microfossils have contributed significantly towards our understanding of Cretaceous and Cenozoic paleoclimate, especially in determining the sea-surface and bottom-water temperatures. There are two approaches to paleotemperature estimation: the known or inferred temperature preferences of microfossils (*biological approach*) and the stable isotopes and trace elements in the carbonate shells of microfossils (*geochemical approach*).

Temperature is an important control on the distribution of organisms in general. In the present day, the distribution of planktic foraminifera is strongly limited by latitude. The cool water is characterized by *Neogloboquadrina pachyderma* and *Globigerina bulloides* and the warm water contains *Globigerinoides sacculifer*, *Sphaeroidinella dehiscens* and *Candeina nitida* (Fig. 11.8). Many of the extant species extend into the geologic past and their present-day distribution can be used with confidence to interpret Neogene and Quaternary paleoclimate. There are well-recognized morphotypic variations in temperature in some species. The high arched aperture *Globigerinoides ruber* occurs in warm water while the low arched aperture occurs in the species inhabiting cold water. The predominantly cool water *Neogloboquadrina pachyderma* shows dextral coiling in relatively warmer waters. The relative abundance of dextral and sinisterly coiled forms is a common tool for

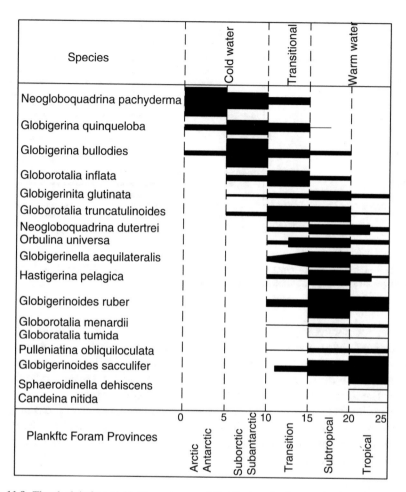

Fig. 11.8 The planktic foraminiferal provinces and distribution of select species in modern oceans (redrawn after Be and Tolderlund 1971, with permission © Cambridge University Press)

inferring glacial–interglacial intervals in the Quaternary period. The larger benthic foraminifera occur in warm, tropical climates and their distribution parallels the distribution of hermatypic corals. The living larger benthic foraminifera occur in the temperature range of ~14–34 °C (Fig. 11.9), but general distribution is limited by winter minimum isotherms between 15 and 20 °C. *Sorites orbiculus* and *Amphistegina* spp. have wider latitudinal variation and can tolerate lower temperatures.

Most paleoclimate reconstructions are based on the oxygen isotopic composition of the carbonate shells of foraminifera. A more reliable estimate of paleotemperature can be achieved by decoupling the ice effect in the oxygen isotopic composition by Mg/Ca ratios in the same phase. The oxygen isotopic data of planktic and deep-sea

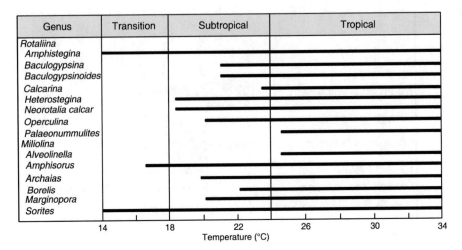

Genus	Transition	Subtropical	Tropical

Rotaliina
Amphistegina
Baculogypsina
Baculogypsinoides
Calcarina
Heterostegina
Neorotalia calcar
Operculina
Palaeonummulites
Miliolina
Alveolinella
Amphisorus
Archaias
Borelis
Marginopora
Sorites

14   18   22   26   30   34
Temperature (°C)

Fig. 11.9 Temperature tolerance of modern larger benthic foraminifera (adapted from Langer and Hottinger 2000, with permission © Micropaleontology Press)

benthic foraminifera, providing temperatures of sea surface and bottom waters, respectively, have given deep insight into climate and ocean circulation in the Cretaceous and Cenozoic. This was possible due to sustained efforts in generating isotope records at higher resolution on the high quality cores recovered in DSDP and ODP. The significant findings were recognition of global climate change, abrupt shifts in climate at the scales of few 100s–1000s k years, and perturbations in global carbon cycles in the Cenozoic. Several issues of these findings are still under debate in paleoclimate research. Some salient findings of the Cenozoic climate and paleo-monsoon are discussed below.

Cenozoic Paleoclimate

A compilation of oxygen and carbon isotopic data of deep-sea benthic foraminifera led to recognition of general trends in Cenozoic climate at a global scale (Fig. 11.10; Zachos et al. 2001). The Cenozoic climate varied from the extremes of warming to the formation of polar ice caps. The total change in $\delta^{18}O$ of deep-sea foraminifera during this era is 5.4‰, of which 3.1‰ is ascribed to deep-sea cooling and the remainder to the growth of ice sheets. The major interpretations are as follows:

The early Cenozoic was marked by a distinct episode of global warming from the Middle Paleocene to Early Eocene, culminating in the Early Eocene Climatic Optimum (EECO). The most pronounced warming occurred at the boundary of the Paleocene and Eocene, known as the Paleocene–Eocene Thermal Maximum (PETM). Deep-sea warming at this time was first recorded by Kennett and Stott (1991) and has been an active area of research since due to a significant rise in tem-

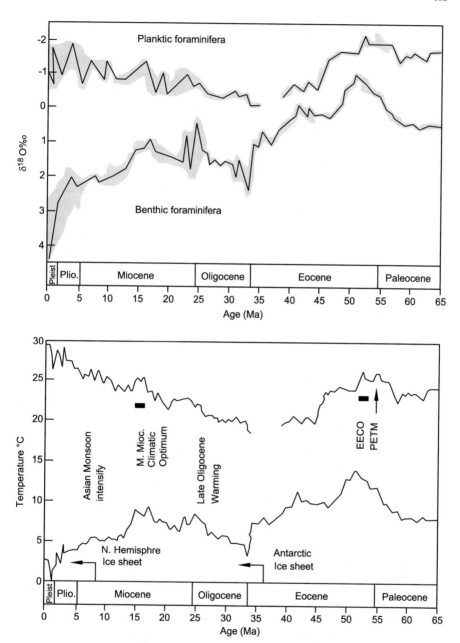

Fig. 11.10 δ¹⁸O of planktic and benthic foraminifera and the estimated temperatures of the surface and bottom waters during the Cenozoic (redrawn after Wright 2009, with permission © Springer Science + Business Media)

perature in a short interval of about 100k years. The PETM records a temperature rise of 5–6 °C and a negative excursion of organic carbon $\delta^{13}C$ of ~3‰ magnitude. The Early Eocene warming was followed by a 17-M-year-long cooling trend until the Early Oligocene, recording a decline in deep-sea temperature of 6 °C. A short interval of warming in the Middle Eocene occurred at ~40 Ma (Bohaty et al. 2009). This event, known as the Middle Eocene Climatic Optimum (MECO), lasted for about 500k years. It is estimated that, in low latitudes, temperatures exceeded 30 °C in the Eocene (Saraswati et al. 1993; Pearson et al. 2007).

Following the cooling and formation of Antarctic continental ice sheets, a warming in the later part of the Oligocene reduced the extent of the Antarctic ice. The warm period peaked in the Middle Miocene Climatic Optimum (MMCO) (17–15 Ma). Two anomalous temperatures are recorded at the Eocene–Oligocene boundary (O_i–1) and the Oligocene–Miocene boundary (M_i–1), respectively, characterized by positive $\delta^{18}O$ excursions, implying cooling in bottom-water temperatures. A gradual cooling followed the MMCO and a permanent ice sheet formed in Antarctica by 10 Ma. A warming trend in the early Pliocene continued until ~3.2 Ma, when abrupt cooling again led to development of ice sheets in the northern hemisphere. The diminishing equatorial circulation and development of permanent ice sheets in the northern hemisphere appears to have been responsible for the establishment of bipolar symmetry in climate change starting three million years ago.

How did the marine ecosystem respond to this dramatic increase in deep-sea temperature during the PETM? Different groups of microfossils responded differently, and the major changes were as follows (McInerney and Wing 2011):

1. There was a marked extinction of benthic foraminifera, reducing 30–50 % of its diversity. They also decreased in size, possibly due to low-oxygen bottom-water conditions.
2. Planktic foraminifera shifted their geographic ranges. For example, *Morozovella*, previously inhabiting tropical water, migrated to high latitudes.
3. Nannoplankton had heterogeneous changes. The rates of both origination and extinction were very high. Their turnover rate was highest in the Cenozoic.
4. In dinoflagellates, the genus *Apectodinium* increased markedly in abundance and geographic range, making it an indicator taxon of the PETM.
5. Ostracoda shows a varied response from decreased diversity and abundance to rapid speciation.

A major interest in Cenozoic paleoclimate in recent years has been focused on knowing the source of carbon supposedly released at the PETM. The destabilization of methane clathrate (the ice crystals with trapped methane, stable in deep-sea sediments) due to a rise in temperature is viewed as the major source. Another view holds thermogenic methane released by injection of magma into organic-rich sediments as a source of carbon at the PETM. The warmth of the PETM and the accompanying carbon isotope excursion (CIE) have emerged as a geological analogue for understanding the greenhouse theory that envisages global warming due to increased CO_2 in the atmosphere from anthropogenic sources.

Table 11.5 Microfossils as tracers of monsoons

Microfossil tracers	Process	Remarks
Foraminifera: *Globigerina bulloides* *Neogloboquadrina dutertrei* *Globigerina falconensis*	Upwelling and productivity: Summer monsoon (India) Winter monsoon (East Asia) Winter monsoon (India)	Abundance is correlated with upwelling/productivity/cooling associated with monsoons
Coccoliths: *Gephyrocapsa oceanica* *Florisphaera profunda* *Emiliania huxleyi* dominated assemblage	Upwelling	High coccolithophore fluxes in summer and winter monsoons
Diatoms	Surface water productivity	Diatom Accumulation Rate (DAR) = diatom abundance × sedimentation rate Higher DAR implies higher fertility
Radiolarians	Surface water productivity	Thermocline Surface Radiolarian (TSR) Index is the ratio of the Total individual of subsurface and intermediate population *and* the Total individual of surface population Higher TSR implies enhanced surface productivity
Dinoflagellates (calcareous cysts)	Upwelling and nutrient	Flux = Absolute cyst abundance × Total mass flux in mg/m^2 d Highest cyst flux in summer monsoon and lowest in inter-monsoon

Compiled after Andruleit et al. 2000; Wang and Abelmann 2002; Wang et al. 2005; Wendler et al. 2002

Paleomonsoon

The monsoon is an important aspect of modern climate, causing cool, dry winters and warm, humid summers over a large stretch of continents, from parts of Africa through India and East Asia to northern Australia. The seasonal change in atmospheric circulation and precipitation also causes seasonal change in strength and direction of ocean currents, sea-surface temperatures and seawater salinity in the Indian Ocean and the South China Sea (Wang et al. 2005). The various manifestations of a monsoon, including the strength and direction of winds, the resultant upwelling in the ocean, degree of precipitation and the consequent weathering of continental rocks, leave their records in oceanic and continental sediments. A number of tracers and proxies developed to interpret the patterns of monsoons in the geologic past helped reconstruct the dynamics of climate change in a robust manner. The majority of reconstructions, particularly of the oceanic records, are microfossil based. The commonly used microfossils include foraminifera, coccoliths, pteropods, ostracoda, radiolarians, diatoms and spore-pollens (Table 11.5).

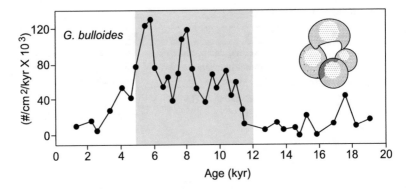

Fig. 11.11 Variation in flux of *Globigerina bulloides* in a sediment core from the Arabian Sea. The high abundance and high flux between 5 and 12 ka suggest upwelling and intensification of the southwest monsoon (simplified after Naidu 2007)

Planktic foraminifera are the most extensively used tracers in marine records due to their rapid response to surface hydrography and monsoon-induced upwelling. The sediment trap and plankton tows data indicate that the subpolar species *Globigerina bulloides* increases in abundance in the regions of monsoon-driven upwelling in low latitudes and, therefore, its census count marks periods of monsoon intensification. The advantages of *G. bulloides* as a monsoon tracer are (1) its unique association with the summer monsoon, (2) linear correlation with the surface cooling due to upwelling, and (3) strong sensitivity to wind speed and the monsoonal atmospheric pressure gradient (Gupta et al. 2003). The upwelling indices, based on planktic foraminifera from cores off Oman, suggest that the southwest monsoon intensified at 12 ka and reached a peak between 10 and 5 ka. It started weakening from 5 ka and the weakest phase was at 3.5 ka. The flux of a number of species of planktic foraminifera increases between 5 and 12 ka, including that of *G. bulloides* (Fig. 11.11; Naidu 2007). The upwelling-induced productivity was higher in the glacial period than in the Holocene. Several studies have attempted to understand the relationship of the Asian monsoon with other elements of global climate. The productivity declines occurred during cold events in the North Atlantic, and it is inferred that monsoon winds (which drive monsoon-induced upwelling) were weak when the North Atlantic was cold (Singh et al. 2011), indicating a link between monsoons and the North Atlantic climate.

The upwelling drives the cold, nutrient-rich bottom water to the surface. This results in a decrease in sea-surface temperature and an increase in nutrient and productivity. These changes captured by the $\delta^{18}O$ and $\delta^{13}C$ of foraminifera complement the census counts by providing input on temperature and productivity changes, respectively. The $\delta^{18}O$ of *Globigerinoides ruber* in sediment traps in the Arabian Sea recorded a 1‰ increase, reflecting about 4 °C cooling of the surface water with upwelling (Curry et al. 1992). Furthermore, the coupled $\delta^{18}O$-Mg/Ca of planktic foraminifera in cores of the Arabian Sea estimates the seasonal temperature range over the past 20k years as fluctuating on an average from 11 °C in the LGM to 13 °C in the Holocene, compared with

Table 11.6 Variability in intensity of monsoons in the past 20k years (western Arabian Sea after Naidu and Malmgren 1996: eastern Arabian Sea, after Sarkar et al. 2000)

Western Arabian Sea	Eastern Arabian Sea
3.5–5.0 ka: Weakening of monsoons	10–2.0 ka: Evaporation—precipitation steadily decreased due to increasing trend in summer monsoon rainfall
5.0–12.0 ka: Intensification of monsoons	
12.0–16.0 ka: Weaker monsoons	
16.0–19.0 ka: Relatively stronger monsoons	

16 °C in the present day (Ganssen et al. 2011). The oxygen isotope data of *Globigerinoides sacculifer* and *Globorotalia menardii* employed to estimate the excess of evaporation over precipitation (E-P) indicates a steadily decreasing value from 10 to 2 ka, likely due to increasing rainfall (Sarkar et al. 2000). Several phases of intensification and weakening in monsoons are recorded in the cores of the Arabian Sea based on planktic foraminifera and isotopic data (Table 11.6).

Most studies on monsoons are for the Quaternary period, although monsoons have a longer history. The tectonic factors believed to have controlled the evolution of the monsoon include (1) uplift of the Tibetan Plateau and the Himalayas, (2) retreat of the Paratethys during the Oligocene to Late Miocene, and (3) closure of the Indonesian seaway, affecting ocean circulation and sea-surface temperatures in the Indian Ocean. Though some studies based on DSDP cores from the northern Indian Ocean have suggested the beginning of the Indian monsoon at about 12–11 Ma, the upwelling indicator taxa in the cores of the Arabian and South China Seas have revealed the onset of the monsoon at about 8.0 Ma (Late Miocene). Long-range stratigraphic records are required to know the complete history of the monsoon and the role of tectonic forcing in its evolution.

References

Andruleit HA, von Rad U, Bruns A, Ittekkot V (2000) Coccolithophore fluxes from sediment traps in the northeastern Arabian Sea off Pakistan. Mar Micropaleontol 38:285–308

Banerji RK (1990) Heavy metals and benthic foraminiferal distribution along Bombay coast, India. In: Takayanagi Y, Saito T (eds) Studies in benthic foraminifera. Tokai University Press, Tokyo, pp 151–157

Be AWH, Tolderlund DS (1971) Distribution and ecology of living planktonic foraminifera in the surface waters of the Atlantic and Indian Oceans. In: Funnell BM, Riedel WR (eds) Micropaleontology of the oceans. Cambridge University Press, London, pp 105–149

Beavington-Penney SJ, Racey A (2004) Ecology of extant nummulitids and other larger benthic foraminifera: applications in palaeoenvironmental analysis. Earth Sci Rev 67:219–265

Bignot G (1985) Elements of micropalaeontology. Graham & Trotman, London

Bohaty SM, Zachos JC, Florindo F, Delaney M (2009) Coupled greenhouse warming and deep-sea acidification in the Middle Eocene. Paleoceanography. doi:10.1029/2008PA001676

Brasier MD (1995) Fossil indicators of nutrient levels. 1: eutrophication and climate change. In: Bosence DWJ, Allison PA (eds) Marine palaeoenvironmental analysis from fossils. Geological Society, London, pp 113–132, Special Publication: 83

Corliss BH, Fois E (1990) Morphotype analysis of deep-sea benthic foraminifera from the north-west Gulf of Mexico. Palaios 5:589–605

Culver SJ (1988) New foraminiferal depth zonation of the northwestern Gulf of Mexico. Palaios 3:69–85

Curry WB, Ostermann DR, Guptha MVS, Ittekkot V (1992) Foraminiferal production and monsoonal upwelling in the Arabian Sea: evidence from sediment traps. In: Summerhayes CP, Prell WL, Emeis KC (eds) Upwelling systems: evolution since the Early Miocene. Geological Society, London, pp 93–106, Special Publication: 64

Fujita K, Hikami M, Suzuki A, Kuroyanagi A, Sakai K, Kawahata H, Nojiri Y (2011) Effects of ocean acidification on calcification of symbiont-bearing reef foraminifers. Biogeosciences 8:2089–2098

Ganssen GM, Peeters FJC, Metcalfe B, Anand P, Jung SJA, Kroon D, Brummer GJA (2011) Quantifying sea surface temperature ranges of the Arabian Sea for the past 20000 years. Clim Past 7:1337–1349

Gooday AJ (1994) The biology of deep-sea foraminifera: a review of some advances and their applications in paleoceanography. Palaios 9:14–31

Gupta AK, Anderson DM, Overpeck JT (2003) Abrupt changes in the Asian southwest monsoon during the Holocene and their links to the North Atlantic Ocean. Nature 421:354–357

Hallock P, Glenn EC (1986) Larger foraminifera: a tool for paleoenvironmental analysis of Cenozoic carbonate depositional facies. Palaios 1:55–64

Hallock P, Lidz BH, Cockey-Burkhard EM, Donnelly KB (2003) Foraminifera as bioindicators in coral reef assessment and monitoring: the FORAM Index. Environ Monit Assess 81:221–238

Honisch B, Ridgwell A, Schmidt DN et al (2012) The geological record of ocean acidification. Science 335:1058–1063

Jones RW, Charnock MA (1985) Morphogroups of agglutinating foraminifera: their life positions and feeding habits and potential applicability in (paleo)ecological studies. Rev Paléobiol 4:311–320

Kaiho K (1994) Benthic foraminiferal dissolved-oxygen index and dissolved-oxygen levels in the modern ocean. Geology 22:719–722

Keen MC (1993) Ostracods as palaeoenvironmental indicators: examples from the Tertiary and Early Cretaceous. In: Jenkins DG (ed) Applied micropalaeontology. Kluwer Academic, The Netherlands, pp 41–67

Kennett JP, Stott LD (1991) Abrupt deep sea warming, paleoceanographic changes and benthic extinctions at the end of the Paleocene. Nature 353:225–229

Langer M, Hottinger L (2000) Biogeography of selected "larger" foraminifera. Micropaleontology 46(Suppl 1):105–127

Martin RE (ed) (2000) Environmental micropaleontology – the application of microfossils to environmental geology. Kluwer, New York

McInerney FA, Wing SL (2011) The Paleocene–Eocene thermal maximum: a perturbation of carbon cycle, climate and biosphere with implications for the future. Annu Rev Earth Planet Sci 39:489–516

Miller KG, Kominz MA, Browning JV, Wright JD, Mountain GS, Katz ME, Sugarman PJ, Cramer BS, Christie-Blick N, Pekar SF (2005) The Phanerozoic record of global sea-level change. Science 310:1293–1298

Miller KG, Mountain GS, Wright JD, Browning JV (2011) A 180-million-year record of sea level and ice volume variations from continental margins and deep sea isotopic records. Oceanography 24(2):40–53

Mojtahid M, Jorissen F, Durrieu J et al (2006) Benthic foraminifera as bioindicators of drill cuttings disposal in tropical East Atlantic outer shelf environments. Mar Micropaleontol 61:58–75

Murray JW (1995) Microfossil indicators of ocean water masses, circulation and climate. In: Bosence DWJ, Allison PA (eds) Marine palaeoenvironmental analysis from fossils. Geological Society, London, pp 245–264, Special Publication: 83

Murray JW (2006) Ecology and applications of benthic foraminifera. Cambridge University Press, Cambridge

Naidu PD (2007) Influence of monsoon upwelling on the planktonic foraminifera off Oman during Late Quaternary. Ind J Mar Sci 36(4):322–331

Naidu PD, Malmgren BA (1996) A high-resolution record of late Quaternary upwelling along the Oman Margin, Arabian Sea based on planktonic foraminifera. Paleoceanography 11(1):129–140

Nixon SW (1995) Coastal marine eutrophication: a definition, social causes and future concerns. Ophelia 41:199–219

Pearson PN, Palmer MR (2000) Atmospheric carbon di-oxide concentrations over the past 60 million years. Nature 406:695–699

Pearson PN, van DongenBE NCJ, Pancost RD, Schouten S, Singano JM, Wade BS (2007) Stable warm tropical climate through the Eocene epoch. Geology 35:211–214

Saraswati PK, Ramesh R, Navada SV (1993) Palaeogene isotopic temperatures in western India. Lethaia 26:89–98

Sarkar A, Ramesh R, Somayajulu BLK, Agnihotri R, Jull AJ, Burr GS (2000) High resolution Holocene monsoon record from the eastern Arabian Sea. Earth Planet Sci Lett 177:209–218

Sen Gupta BK, Machain-Castillo ML (1993) Benthic foraminifera in oxygen-poor habitats. Mar Micropaleontol 20:183–201

Singh AD., Jung SJA, Darling K, Ganeshram R, Ivanochko T, Kroon D (2011) Productivity collapse in the Arabian Sea during glacial cold phases. Paleoceanography 26. doi: 10.1029/2009 PA 001923

Van der Zwaan GJ, Jorissen FJ, de Stigter HC (1990) The depth dependency of planktonic/benthic foraminiferal ratios: constraints and applications. Mar Geol 95:1–16

Wang P, Clemens S, Beaufort L, Braconnot P, Ganssen G, Jian Z, Kershaw P, Sarnthein M (2005) Evolution and variability of the Asian monsoon system: state of the art and outstanding issues. Quat Sci Rev 24:595–629

Wang R, Abelmann A (2002) Radiolarian responses to paleoceanographic events of the southern South China Sea during the Pleistocene. Mar Micropaleontol 46:25–44

Wendler I, Zonneveld KAF, Willems H (2002) Production of calcareous dinoflagellate cysts in response to monsoon forcing off Somalia: a sediment trap study. Mar Micropaleontol 46:1–11

Wright JD (2009) Cenozoic climate change. In: Gornitz V (ed) Encyclopedia of paleoclimatology and ancient environments. Springer, The Netherlands, pp 148–155

Yanko V, Ahmad M, Kaminski M (1998) Morphological deformities of benthic foraminiferal tests in response to pollution by heavy metals: Implications for pollution monitoring. J Foraminifer Res 28:177–200

Zachos JC, Pagani M, Sloan L, Thomas E, Billups K (2001) Trends, rhythms and aberrations in global climate 65 Ma to present. Science 292:686–693

Zachos JC, Rohl U, Schellenberg SA, Sluijs A, Hoddell DA, Kelley DC, Thomas E, Nicolo M, Raffi I, Lourens LJ, McCarren H, Kroon D (2005) Rapid acidification of the ocean during the Paleocene–Eocene thermal maximum. Science 308:1611–1615

Zeebe RE, Zachos JC, Caldeira K, Tyrrell T (2008) Carbon emissions and acidification. Science 321:51–52

Further Reading

Armstrong HA, Brasier MD (2005) Microfossils, IIth edn. Blackwell, Oxford

Bignot G (1985) Elements of micropalaeontology. Graham & Trotman, London

Haq BU, Boersma A (1978) Introduction to marine micropalaeontology. Elsevier, New York

Chapter 12
Basin Analysis and Hydrocarbon Exploration

12.1 Introduction

Sedimentary basins are host to mineral and fuel resources, the exploration of which requires a sound knowledge of the evolution of the sedimentary basin that hosts them. Basin analysis traces the evolution of the sedimentary basin and includes the origin, stratigraphic architecture, paleogeography and subsidence history of the basin. Micropaleontology is applied at various stages of basin analysis. The stratigraphic framework of the basin is primarily based on micropaleontology and key data in reconstruction of paleogeography and subsidence history are provided by microfossils. Furthermore, two modern techniques in basin analysis, the seismic stratigraphy and sequence stratigraphy, need micropaleontologic support in a major way. It was a revolutionary concept that seismic reflections are essentially timelines (Vail et al. 1977), but the reflections are required to be biostratigraphically calibrated. In sequence stratigraphy, microfossils contribute significantly to the recognition of sequence boundaries, transgressive surface, maximum flooding surface and systems tracts. Micropaleontology is indispensible wherever information on age and environment of the sedimentary fill of the basin is required in basin analysis.

Historically, the petroleum industry nurtured micropaleontology for its outstanding contributions in subsurface exploration. Microfossils, due to their small size, abundance and occurrence in sediments of all environments, proved useful in the limited size of the core and drill cuttings. They have been found useful in all three phases of hydrocarbon exploration, including exploration for prospects, appraisal of discoveries and development of fields. The role of micropaleontology has expanded beyond its traditional use in determining the age of strata and correlation of wells. Biostratigraphy is being used successfully in the monitoring and

© Springer International Publishing Switzerland 2016
P.K. Saraswati, M.S. Srinivasan, *Micropaleontology*,
DOI 10.1007/978-3-319-14574-7_12

guiding of horizontal or deviated drilling through reservoirs, known as "biosteering". This chapter discusses the application of micropaleontology in basin analysis and hydrocarbon exploration.

12.2   Calibration of Seismic Sections

Seismic reflections are due to a contrast in acoustic impedance (product of velocity of seismic wave and density of the rock) across interfaces. The interfaces include bedding planes, unconformities and pore fluids. The seismic sections are divided into depositional sequences based on the presence of unconformities. Unconformities generate reflections because they separate beds with different lithologies and, hence, of different acoustic impedance. Unconformity, however, need not be reflective, because it depends on the acoustic impedance contrast. For example, unconformity between two parallel beds of limestone (paraconformity) may have very weak reflection. On the other hand, a conformable sequence of limestone and chert may give strong reflection due to density contrast. Therefore, although seismic reflections are timelines, they require biostratigraphical calibration. Moreover, the reflections generated by unconformities represent a hiatus, and its magnitude is likely to differ from one place to another in the basin. In a transgressive sequence, for example, when progressively younger sediments overlie the basement, the hiatuses are of different magnitudes in different parts of the basin. The seismic sections cannot reveal this unless biostratigraphically calibrated. The seismic attributes, such as amplitude and continuity, related to depositional conditions are calibrated through geological data from wells.

In seismic stratigraphy, the bounding unconformities on seismic sections are delineated, and then chronostratigraphy of the basin is reconstructed as follows (Veeken 2007):

1. The systems tracts are established in the depositional sequences.
2. Various seismic facies are outlined.
3. The depositional environment and distribution of gross lithological units are interpreted by incorporating all available geological controls from the well data.
4. A stratigraphic chart with some arbitrary time scale that gives an overview of the distribution of time-equivalent deposits in the basin and the relative importance of various unconformities is prepared.

In the stratigraphic chart, the seismically defined time units are calibrated by the biostratigraphic data generated on the drilled wells in the basin (Fig. 12.1). It assigns a precise time to specific intervals of the depositional sequence and presents a chronostratigraphic chart of the basin. The architecture of the basin is more proportional to the actual elapsed time and hiatuses are accurately represented in the biostratigraphically constrained chronostratigraphic chart.

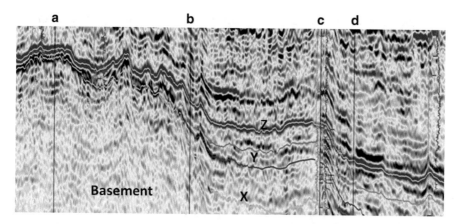

Fig. 12.1 A seismic section showing the basement and the overlying sediments in a basin. *A–D* mark the positions of the wells and *X, Y, Z* are some of the interpreted seismic markers. The markers are calibrated biostratigraphically

12.3   Sequence Stratigraphy

Peter Vail and his group at the Exxon Corporation gave a new lease to the concept of recognizing stratigraphic units on the basis of bounding unconformities. Several other previous researches had used this concept, the most notable of which was recognition of six unconformity bound sequences in North American sedimentary succession by Sloss (1963). A novelty of Vail's group was in demonstrating that sequences could be recognized on seismic reflections and, thus, tracing of sequence architecture in the subsurface and on the basinal scale was made possible. Emery and Myers (1996), Catuneanu (2006) and Miall (2010) made an extensive review on the development of sequence stratigraphy, the limitations of the methodology and a way forward in this science. In the following overview of sequence stratigraphy, the definitions of sequence-related terms are largely from Miall (2010) and the micropaleontological characteristics are mainly those listed by Olson and Thompson (2005).

A depositional sequence is a stratigraphic unit, composed of a relatively conformable succession of genetically related strata, bounded at its top and base by unconformities or their correlative conformities. The unconformities in the definition refer to those developed due to subaerial exposure, and when traced laterally into deposits of deep marine environment, grade to correlative conformity. The three key stratal surfaces in depositional sequences and their micropaleontologic characteristics are explained below.

Sequence boundary: This forms in response to a relative fall in the sea level. It is
    defined by an unconformity up-dip and correlative conformity down-dip. The

sequence boundary is readily recognized by a contrast in the age and paleoenvironment of the microfossils below and above the boundary. It is characterized by (1) abrupt truncation or decrease in abundance and diversity of marine microfossils, (2) upward increase in terrestrial spores and pollens, (3) decrease in ratio of planktic and benthic foraminifera (P/B), and (4) increase in reworked microfossils. The evidence of reworking includes the presence of microfossils of significantly different environments, different ages and different states of preservation. In sequence stratigraphy of carbonate platforms, when sea-level fall exceeds subsidence rates, the exposed carbonates are subject to karstification. Sequence boundary in this setting is identified by karstic surface. A rapid transgression, high nutrient levels or high siliciclastic input may stop carbonate production, forming *drowning unconformity*.

Transgressive surface (or *marine flooding surface*): This represents the first major flooding surface following the sequence boundary. It separates older strata from younger across which there is evidence of an abrupt increase in water depth. It is marked by an increase in marine microfossils.

Maximum flooding surface: This marks the deepest water facies in the sequence and separates the transgressive unit below and the regressive unit above. It is characterized by (1) peak abundance of planktic microfossils, including planktic foraminifera, calcareous nannofossils and dinocysts, (2) deep water benthic microfossils, (3) dominance of low-oxygen-tolerant benthic foraminifera and infaunal foraminifera, (4) minimum terrestrial spores and pollens, (5) minimum reworked microfossils, and (6) the maximum total foraminiferal number (TFN) and P/B ratio.

The sequence development can be explained in terms of base-level cycles. The base level refers to the level above which erosion will occur, controlled by the sea level on the continental margin. The change in sea level may be due to eustasy and subsidence/uplift of the sea floor. It is difficult to distinguish the two processes and, therefore, relative sea level is generally used. The positions of the generation of the major surfaces and systems tracts in relation to the sea-level curve are shown in Fig. 12.2 and described below.

Highstand Systems Tract (HST): This overlies the MFS and is overlain by the sequence boundary. It forms when base-level rise is at its highest point and in the process of slowing down prior to commencement of slow fall. The micropaleontological characteristics of HST are (1) shallowing-up benthic foraminiferal assemblage, (2) gradual upward reduction in open ocean planktic microfossils, (3) upward decrease in P/B ratio, (4) irregular last occurrences (LOs) due to a problem of ecology and clastic dilution, (5) gradual upward increase in derived terrestrial fossils, and (6) diachronous biofacies boundaries.

Falling Stage Systems Tract (FSST): This occurs during the sea-level cycle when the sediment supply on the continental shelf and continental slope is at its maximum. A fall in the base level exposes the coastal plain and the continental shelf

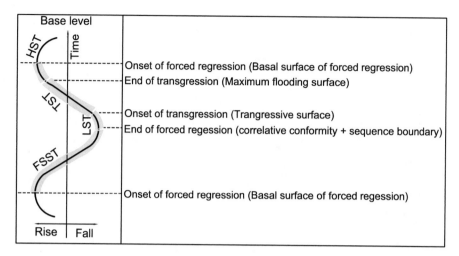

Fig. 12.2 The major surfaces and systems tracts in relation to base-level cycles

to subaerial erosion. The FSST is characterized by derived terrestrial fossils and reworked microfossils of the previously deposited HST.

Lowstand Systems Tract (LST): This occurs when the sea level is at its bottom and the depositional trend shifts from seaward to landward and constitutes a small part of the sequence. The micropaleontologic characteristic of LST is in most ways similar to that of the sequence boundary.

Transgressive Systems Tract (TST): This is formed during a rise in the base level when that rise exceeds sediment supply. It leads to retrogradation of depositional systems. The micropaleontologic characteristics of TST include (1) gradual upward decrease in spores and pollens, (2) upward increase in planktic foraminifera and calcareous nannofossils, (3) upward deepening indicated by benthic foraminiferal assemblage, (4) increase in P/B ratio, (5) upward decrease in reworked microfossils, and (6) a diachronous biofacies boundary.

12.4 Subsidence Analysis

The estimation of subsidence of sediments is an integral part of basin analysis and of crucial importance in thermal maturation of organic matter for hydrocarbon generation. The total subsidence is the combined effect of sediment load and tectonic subsidence of the basin. Subsidence analysis (or geohistory analysis) decouples the effects of load and tectonics through a procedure called *backstripping*. This involves progressive removal of sedimentary load from a basin, correcting for compaction,

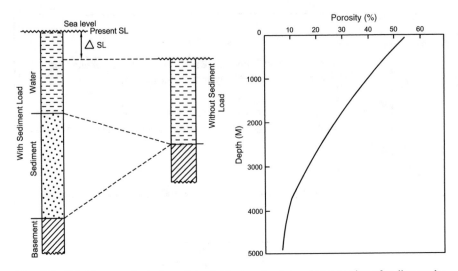

Fig. 12.3 Subsidence analysis through backstripping, involving decompaction of sediments by the porosity curve (*right*) and removal of sediment load (*left*) for each increment of time (redrawn after Steckler and Watts 1978, with permission © Elsevier)

paleobathymetry and changes in sea level and calculating the depth to the basement. A summary of the method is given below and for detailed methodology, readers are referred to Allen and Allen (2005).

1. The stratigraphic column is divided into finer time slices, the ages of which are accurately determined by biostratigraphy, aided by magnetostratigraphy and isotope stratigraphy wherever possible.

2. The time slices are decompacted to their original thickness by measuring changes in porosity with depth. The porosity is measured on core plugs or it is estimated from petrophysical logs. The compaction is lithology dependent, being high in shale and low in sandstone and limestone. The decompacted layers are added one by one to the basement (Fig. 12.3).

3. The decompacted thicknesses are subject to corrections for water depth and eustasy, so as to know the compaction and tectonic contributions.

4. The corrections for water depth are done by interpreting the paleobathymetry of the time slices recognized above. All the evidence of paleobathymetry is considered to constrain water depth as accurately as possible, but it is well known that microfossils, particularly benthic foraminifera, are the best indicators of water depths in marine succession.

5. Eustatic change causes isostatic compensation. There is no consensus at present on the global eustatic curve to make corrections in decompacted thicknesses for backstripping. It is ignored as a first approximation. The total subsidence is, thus, due to sediment/water load and tectonic force.

6. The isostatic effect of sediment load is removed from the total subsidence to obtain the tectonic subsidence of the basement. This is done through backstripping using Airy isostasy (for an exercise on backstripping, see: www.erdw.ethz.ch/Allen).

12.5 Thermal History

For an exploration geologist, the subsidence history of a basin is important from the point of view of organic matter maturation. As sediments are buried through depths, they pass through different regimes of temperature and pressure. The cooking of organic matter to produce hydrocarbon is temperature dependent (Fig. 12.4). There are biological indices, such as vitrinite reflectance of dispersed organic matter, the spore colour index and the acritarch alteration index used by petroleum geologists, to assess the thermal maturity of the source rocks and the *oil window zone* (Fig. 12.4). The type I and II source rocks become mature for generation of oil when the spore colour index ranges from 3 to 6. The vitrinite reflectance values are superimposed on the subsidence curve to decipher the time of formation of hydrocarbon in the basin.

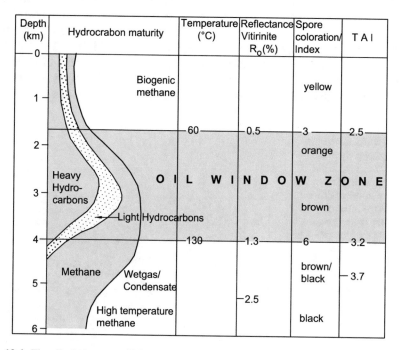

Fig. 12.4 The oil window zone and the biological indicators of the thermal maturity of the source rocks

12.6 Biosteering

Micropaleontologists have a well-defined role in the exploration of prospects. In the wildcat phase of exploration, a well is drilled on a structure recognized on a seismic section without detailed stratigraphic information. As soon as the well is drilled, a micropaleontologist is required for the purpose of providing a detailed report, including age and paleoenvironment of the drilled section, and calibrating the seismic section with well stratigraphy. The task of the micropaleontologist also includes selecting points for coring, making decisions on terminal depths, and monitoring the well stratigraphically while drilling is in progress. The micropaleontological data is generated further during the appraisal of discoveries to correlate the appraisal wells by biostratigraphy, along with wire line logs and seismic data. This is an important phase of exploration in which distribution of reservoirs is established to estimate the reserve. The field development starts after the decision is made to produce from the field. Micropaleontologic data is routinely generated on all the wells drilled during the development phase. It further refines understanding of the lateral and vertical distribution of reservoirs. Today, micropaleontologists are using microfossil data in a novel way to guide the drilling of deviated wells through biosteering (Fig. 12.5).

In modern-day exploration, penetrating the reservoir through directional drilling economizes the cost. In horizontal or directional drilling, a well is steered biostratigraphically in a process known as *biosteering*. This involves real-time monitoring of the stratigraphic position relative to the reservoir in a (deviated) well through biostratigraphic techniques applied at the well site (Jones et al. 2005). Biosteering requires an intensive understanding of reservoir biofacies and needs to be integrated with other sedimentologic and petrophysical properties to subdivide

Fig. 12.5 Biosteering to drill horizontally in the reservoir and guide the driller to deviate the well for the purpose of drilling through the reservoir across the fault plane. *A–E* represent the biozones. The biozone *B*, characterized by a specific microfaunal assemblage, helps to constrain the reservoir interval

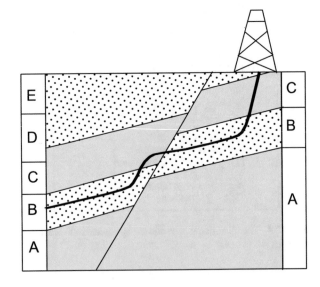

the reservoir units into several zones. Micropaleontologists may use abundance and diversity of microfauna, morphogroup, peak occurrence, multiple microfossils and other such attributes specific to a reservoir to divide it and trace it laterally to guide the drilling. Experience, novelty and an analytical approach are key to its success. Jones (2014) has discussed several case studies from the shallow marine carbonate reservoir of the United Arab Emirates and the deep marine carbonate and siliciclastic reservoirs of the North Sea. In each of these settings, the foraminifera- and nannoplankton-based biostratigraphy enabled the reservoir interval to be divided into several zones and helped in drilling horizontally through the reservoirs. The method has been specifically useful for thinner pay intervals and has saved the industry millions of dollars by optimizing hydrocarbon recovery with a reduced number of wells for production.

References

Allen PA, Allen JR (2005) Basin analysis principles and applications, IIth edn. Blackwell, Oxford

Catuneanu O (2006) Principles of sequence stratigraphy. Elsevier, Amsterdam

Emery D, Myers KJ (1996) Sequence stratigraphy. Blackwell, Oxford

Jones RW, Lowe S, Milner P, Heavey P, Payne S, Ewen D (2005) The role and value of "Biosteering" in hydrocarbon reservoir exploitation. In: Koutsoukos EAM (ed) Applied stratigraphy. Springer, The Netherlands, pp 339–355

Jones RW (2014) Foraminifera and their applications. Cambridge University Press, Cambridge

Miall AD (2010) The geology of stratigraphic sequences, IIth edn. Springer, Heidelberg

Olson HC, Thompson PR (2005) Sequence biostratigraphy with examples from the Plio-Pleistocene and Quaternary. In: Koutsoukos EAM (ed) Applied stratigraphy. Springer, The Netherlands, pp 227–247

Sloss LL (1963) Sequences in the cratonic interior of North America. Geol Soc Am Bull 74:93–113

Steckler MS, Watts AB (1978) Subsidence of the Atlantic type continental margin off New York. Earth Planet Sci Lett 41:1–13

Vail PR, Mitchum RM, Todd RG, Widner JM, Thompson S, Sangree JB, Bubb JN, Hatfield WG (1977) Seismic stratigraphy and global changes in sea level. In: Payton CE (ed) Seismic stratigraphy: application to hydrocarbon exploration, vol 26, AAPG memoir. AAPG, Tulsa, pp 49–212

Veeken PCH (2007) Seismic stratigraphy basin analysis and reservoir characterization. Elsevier, Amsterdam

Further Reading

Jones RW (1996) Micropalaeontology in petroleum exploration. Clarendon, Oxford

Chapter 13
Paleoceanography

13.1 Introduction

Paleoceanography is the study of the evolutionary development of ocean systems through geologic time. It is an interdisciplinary field, integrating stratigraphy, sedimentology, marine micropaleontology, geochemistry and geophysics, and relating them to physical, chemical and biological oceanography. Paleoceanography is one of the youngest branches of earth system science, largely born of the Deep Sea Drilling Project (DSDP), and continues to be nourished by deep-sea exploration. Oceans cover 70.8 % of the earth's surface. The four oceans include the Pacific, Atlantic, Indian, and Southern Ocean. In addition, there is a large number of smaller water bodies connected to these oceans, often referred to as seas, for example, the Arabian Sea, Red Sea, and Persian Gulf, all connected to the Indian Ocean. The depth of the ocean floors is variable, the maximum reaching 10,924 m in the Pacific. Ocean water properties vary aerially and through the depths. High salinity waters are found in enclosed seas and tropical–subtropical regions where evaporation exceeds precipitation and generally decrease poleward. Similarly, the warmest temperatures are found in the tropical regions of the Indian and Pacific Oceans and decrease poleward. Salinity, density and temperature also change vertically and, based on trends in variation, three depth zones are recognized. The upper ~100 m is a mixed layer or surface zone, characterized by water with higher temperature, lower density and lower salinity. There is a rapid change in temperature, density and salinity of water below the mixed zone, continuing for ~1000 m, called the thermocline, pycnocline and halocline, respectively (Fig. 13.1). The lower layer, called the deep zone, contains most of the ocean water column, and

© Springer International Publishing Switzerland 2016
P.K. Saraswati, M.S. Srinivasan, *Micropaleontology*,
DOI 10.1007/978-3-319-14574-7_13

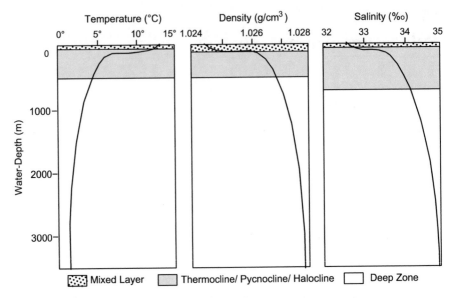

Fig. 13.1 A general profile of variation in temperature, salinity and density of the seawater with depth

properties of the water change slowly in this zone. Thus, the mixed layer and the deep zone are relatively uniform in temperature, salinity and density, while the thermocline and the corresponding pycnocline and halocline are transitional layers between the two.

The wind flowing over the surface slowly drags the water and produces surface ocean currents in the upper 50–100 m of the water column. Below this depth, there are deep-water currents driven by temperature and salinity, known as thermohaline circulation. Two factors, Coriolis force and Ekman transport, affect the direction of ocean currents. The Coriolis force results due to angular momentum of the rotating earth and it causes ocean current to deflect to the right in the northern hemisphere and to the left in the southern hemisphere. The direction of the current is also a balance between the wind and Coriolis forces. The effect of wind decreases and the influence of Coriolis force increases with depth, owing to which the successively deeper layers in the ocean are shifted further to the right, forming a spiral. The average flow over the spiral is the Ekman transport. The net water moment resulting from this is directed 90° in the direction of the wind. There are major current systems in the world oceans (Fig. 13.2), each of which is part of a larger sub-circular

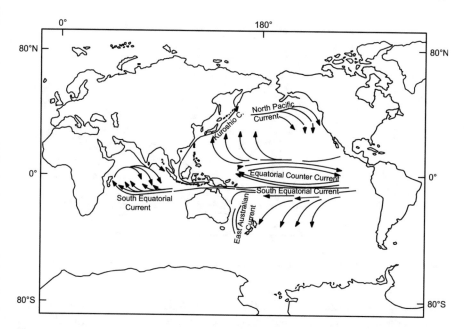

Fig. 13.2 Surface ocean currents of the Indian and Pacific Oceans (redrawn after Srinivasan and Sinha (2000), with permission © Indian Academy of Sciences)

current system, called a gyre (Fig. 13.3). Besides the surface currents, the oceans have deep-water thermohaline circulation (Fig. 13.4) caused by the sinking of dense, cold and saline surface water. In the North Atlantic, the cool and more saline surface water descends to depths of several kilometres, forming the North Atlantic Deep Water (NADW). It spreads to the South Atlantic and flows over the denser Antarctic Bottom Water (ABW), which forms adjacent to the Antarctic continent.

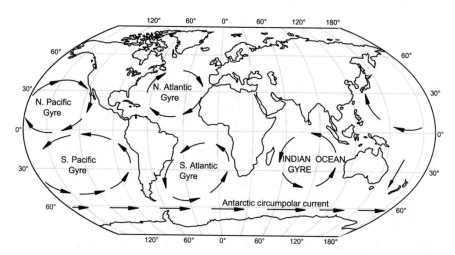

Fig. 13.3 Five major ocean gyres constituted of sub-circular ocean currents that circulate in a clockwise direction in the northern hemisphere and counter-clockwise in the southern hemisphere

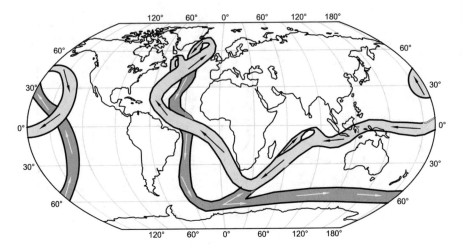

Fig. 13.4 Thermohaline circulation driven by temperature and salinity of the ocean water

The ABW flows into the North Atlantic as the Antarctic Intermediate Water (AAIW) at mean depths of 1 km. The readers are referred to Skinner and Murck (2011) for a detailed description of ocean currents.

From a micropaleontologic perspective, it is of interest that a vast area of the world's ocean floor is covered by sediments called *ooze*, which are composed of more than 30 % of the shells of single-celled planktic and benthic organisms. There are calcareous and siliceous oozes, depending on the chemical compositions of the major contributing shells. The siliceous oozes occur at high latitudes in the Pacific and Indian Oceans. The oceanic sediments also include clay, glacial marine sediment and volcanogenic sediments. The oceanic sediments are the records of past ocean changes, and paleoceanographers use them to reconstruct the history of ancient oceans through biological and chemical proxies.

13.2   Deep-Sea Exploration

Deep-sea exploration represents one of earth science's most successful long-ranging international collaboration programs. The piston-coring technique of the 1940s enabled research vessels to recover long sediment cores from the seafloor. Piston cores from the legendary core cruise of the Swedish Deep-sea Expedition "*Albatross*" in the 1950s greatly enhanced our understanding of ocean bottom sediments. In 1961, when the Dynamic Positioning System (DPS) was successfully developed to keep the drilling platform *CUSS* in position against powerful currents, scientific ocean drilling came into existence as a feasible technology for exploring submarine geology.

Perhaps the most outstanding event in paleoceanography was the advent of the Deep Sea Drilling Project (DSDP) in 1968, using the drilling vessel *Glomar Challenger*. This research vessel not only conducted drilling operations in the world's oceans but also facilitated advanced drilling techniques for recovering undisturbed cores from deep ocean basins. In 1985, the *Glomar Challenger* was replaced by a more sophisticated research vessel, the *JOIDES Resolution*, and a new program, the Ocean Drilling Program (ODP), commenced. The ODP, an international cooperative effort to explore the structure and evolution of ocean basins, conducted 110 expeditions and drilled sites in almost all parts of the world's oceans.

The tremendous scientific success of the DSDP and ODP led to the Integrated Ocean Drilling Program (IODP 2003–2013). A new Japanese deep-sea drilling vessel, the *Chikyu*, equipped with multiple drilling platforms, brought to light enormous deep-sea data hitherto unknown. Starting in October 2013, the IODP participating countries continued their collaborative efforts through the *International Ocean Discovery Program: Exploring the Earth Under the Sea*. Thus, these deep-sea exploration programs over the past four decades have provided an unprecedented opportunity for examination of deep-sea cores, from tropics to poles and from the Mesozoic to the present day. The deep-sea sections provided the scope needed to integrate several stratigraphic techniques and apply them to paleoceanographic reconstructions.

13.3 Approaches to Paleoceanography

Paleoceanographers use a variety of evidence preserved in oceanic sediments to reconstruct the physical, chemical and biological changes in the oceans. The direct evidences are "tracers", defined as objects related to environmental conditions or processes that may provide quantitative or qualitative insight into such conditions or processes. The indirect evidences that provide quantitative values with estimates of uncertainties of the reconstructed parameters are "proxies" (Hillaire-Marcel and de Vernal 2007). There are a number of groups of tracers and proxies of potential use in paleoceanography (Table 13.1). For a reliable reconstruction of past oceans, multiple proxies are used in combination due to the inherent limitations of proxies and the complexity of ocean dynamics. In estimating seawater temperatures, for example, microfossil assemblages, oxygen isotopes and Mg/Ca ratios of the shell carbonates in the same samples will produce a better, more robust interpretation. Microfossils provide a qualitative to semi-quantitative estimate of ocean water temperature, while oxygen isotopic ratios and Mg/Ca ratios of calcareous tests give quantitative estimates of same.

Approaches to paleoceanographic reconstructions employing oceanic microfossils are mainly twofold—one based on direct study of microfossil assemblages (as with tracers) and the other based on the biogeochemistry of their tests (as with proxies). Of the many types of oceanic microfossils, the foraminifera are the best-known group whose modern biogeography, as well as response to changes in the surface and deep waters across the latitudes, is well established. Hence, this group has been widely used for paleoceanographic reconstructions. The most conventional methods adopted include distribution and variation in foraminiferal abundance, benthic/planktic ratios, changes in coiling direction and trace elements, and stable isotopic composition of foraminiferal tests (Table 13.2; see Chap. 11 for details). The surface ultrastructures of some planktic foraminifera provided new insights into paleoenvironmental reconstructions. The SEM studies have revealed

Table 13.1 The most widely used tracers and proxies and their use in interpretation of variables of paleoceanographic significance (compiled from Meissner et al. 2009)

Groups of tracers/ proxies	Examples	Interpreted variables
Microfossil assemblages	Foraminifera, Coccoliths, Radiolarians, Diatoms	Temperature, productivity, water depths
Stable isotopes	$\delta^{18}O$ of foraminifera tests and organic matter	Temperature and extent of continental ice sheets
	$\delta^{11}B$	pH
	$\delta^{15}N$, $\delta^{13}C$	Productivity, nutrient, circulation
Radiogenic isotopes	U and its decay product Th and Pa	Paleoflux and paleocirculation
	^{14}C	Age difference between surface and deep waters
Biogenic compounds	Organic carbon, $CaCO_3$, opal	Productivity, CCD
Elements	Mg/Ca, Sr/Ca in biogenic carbonates	Temperature
	Cd/Ca	Nutrient
	Ba and Ba/Ca	Productivity and alkalinity
Sedimentology	Grain size distribution	Bottom-water current speed (qualitative)
	Rhythmites	Tides
	Mineralogy	Source and direction of sediment transport

that the surface ultrastructures change consistently with changing latitudes and reflect a close relationship with water masses (Srinivasan and Kennett 1975). Vincent and Berger (1981) conducted good reviews on planktic foraminifera, and Gooday (2003) and Jorissen et al. (2007) reviewed benthic foraminifera as tracers of paleoceanography. Employing these techniques, it is now possible to trace marine biogeographic development, changing paleotemperature patterns, fluctuation in carbonate compensation depth (CCD), changes in organic productivity, and opening and closing of ocean gateways and their impact on ocean circulation and climate.

A quantitative approach called transfer functions has gained importance in paleoceanography since the classic work by Imbrie and Kipp (1971) on reconstruction of sea-surface temperatures of the Last Glacial Maximum (LGM). As part of the Climate Long-range Investigation, Mapping And Prediction (CLIMAP) project, the sea-surface temperature pattern of the world's oceans during the LGM as one of the boundary conditions for simulating the atmosphere had to be obtained. Transfer function was applied to estimate temperatures based on planktic microfossils. Another quantitative approach is the modern-analogues technique (MAT), in which the microfossil assemblages of core samples are statistically compared by distance or similarity measures with a number of modern or surface samples. The smaller the distance (or greater the similarity) coefficients between the two samples, the greater the degree of analogy. The basic assumptions in the transfer function method are (1) the core-top fauna is related to hydrographic conditions and (2) the ecology of the species has not changed from the present to the time of the analysed fauna. Guiot and de Vernal (2007) have explained the use of microfossil-based transfer functions in paleoceanography.

Table 13.2 Foraminifera as tracers of paleoceanography

Parameters	Foraminifera or their characteristics	Remarks	Reference
Sea-surface temperature	Coiling direction of *Neogloboquadrina pachyderma*	Sinistral in cold temperature	Bandy (1960)
	Coiling in *G. bulloides*	Sinistral in colder temperature or higher fertility	Naidu and Malmgren (1996)
	Transfer function based on planktic foraminifera	$T = \Sigma\left(P_i \times T_i\right)/\Sigma P_i$ P_i is proportion of species i and T_i is optimal temperature for species i	Berger (1969)
Paleoproductivity	Benthic Foraminiferal Accumulation Rate (BFAR)	$PP = Z.\left(31+1.06\ BFAR\right)/$ $100.\left(k+rz^{0.5}\right)$ PP is primary productivity (g $Cm^{-2}years^{-1}$); Z is water depth in m and $BFAR$ is number of benthic foraminifera $cm^{-2}\ ka^{-1}$	Herguera and Berger (1991)
	Count of *Globigerina bulloides*	High counts in high productivity and upwelling areas	Gupta et al. (2003)
	Bulimina arabiensis	Abundance correlated with productivity and bottom-water oxygenation	Bharti and Singh (2013)
Bottom-water oxygenation	Low diversity and high dominance of *Bolivina, Bulimina, Globobulimina, Uvigerina, Cassidulina* in bathyal oxygen minimum zone (OMZ)	There are no modern foraminifera whose mere presence will indicate low-oxygen bottom water	Sen Gupta and Machain-Castillo (1993)
	Transfer function based on relative frequency of indicator foraminifera of oxic and dysoxic conditions (see original paper for list of taxa)	Foraminiferal oxygen index, $\left[\{O/(O+D)\}\times 100\right]$ where O and D are the number of specimens of oxic and dysoxic foraminifera, respectively	Kaiho (1994)
	Morphologic characteristics	Thin wall, small test, high test porosity	Kaiho (1994); Sen Gupta and Machain-Castillo (1993)
Opening and closing of ocean gateways	Planktic foraminifer *P. spectabilis*	Paleobiogeographic distribution confined to tropical Pacific only during 5.6–4.2 Ma	Srinivasan and Sinha (2000)

Studies from deep-sea cores suggest that the living bathyal–abyssal benthic foraminifera developed in the Middle Miocene about 14 million years ago and, since then, the Neogene benthic foraminiferal assemblages underwent little change (Woodruff 1985). This has provided important clues for understanding the evolution of bottom-water paleoceanography since the Middle Miocene. The uncertainty in the assumption that the ecology of fauna has not changed with time increases for samples older than the Neogene. Due to the complexity of ecological parameters and limitations of practically every proxy data, multi-proxy reconstructions are always preferred for robust interpretation.

13.4   Microfossils in Tracing the History of Cenozoic Oceans

Deep-sea drilling had a major impact on micropaleontology. This can be gauged from the fact that only three major microfossil groups (foraminifera, calcareous nannofossil and radiolarians) were reported in the Initial Reports of the Deep Sea Drilling Project in 1968, rising to 15 microfossil groups by the end of 1978. The deep-sea cores enabled the expansion of micropaleontology in three major areas: oceanic biostratigraphy, paleoenvironmental distribution of microfossils and geochemistry of the microfossil shells (mainly the stable isotopes and trace elements of foraminifera). Paleoceanography, immensely benefited by the progress of oceanic micropaleontology and microfossils, became indispensible in tracing the evolutionary history of oceans. The major highlights in the evolution of oceans, for which micropaleontology played an important interpretational role, are given below.

Glacial–Interglacial Ocean Reconstruction

The seminal work by Emiliani (1955) on the cores of the Caribbean and North Atlantic revealed cycles of oxygen isotopes measured in planktic foraminifera for the first time. These cycles correspond to the cycles of sea-surface temperatures in the Pleistocene and are attributed to warm and cold temperatures. In the general trend of cooling in the Cenozoic, the Pleistocene marks cycles of glacial–interglacial intervals. The land records of glacial periods have poor potential for preservation. The deep-sea cores provided an enormous opportunity to understand how the climate, the oceans and the biota responded to the glacial–interglacial switchovers in the Pleistocene. The CLIMAP project, referred to above, took the task of reconstructing the glacial ocean during the last glacial maximum in 18,000 BP. The biological transfer functions and oxygen isotope composition of planktic foraminifera were employed to estimate sea-surface temperatures for the radiocarbon dated 18,000 BP cores over the world's oceans (CLIMAP Project Members 1976). A map of sea-surface temperatures for a typical summer month during maximum glaciation, called the 18K Map, was prepared (Fig. 13.5). The major conclusions of the project were as follows:

1. There were extensive ice complexes in the Northern Hemisphere and the sea ice in the Southern Hemisphere was significantly greater than it is today.

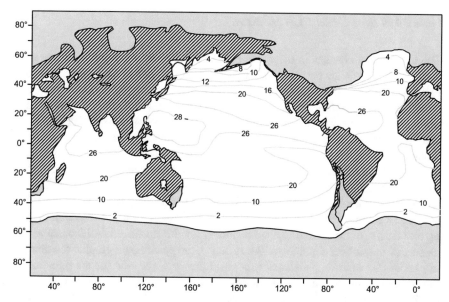

Fig. 13.5 A simplified map to show sea-surface temperatures and extent of ice sheets during the Last Glacial Maximum (LGM) at 18 ka (simplified from CLIMAP (1976), with permission © American Association for the Advancement of Science)

2. Ocean surface was cooler than today by 2.3 °C, on average. There was, however, a marked steepening of thermal gradients along the North Atlantic and Antarctic. The largest change in temperatures was in the North Atlantic Ocean, polewards of 40° N.
3. The temperatures over large areas of the subtropical Atlantic, Pacific and Indian Oceans remained stable, and cooling was <2 °C than it is today.
4. Due to steepened thermal gradient and, as a consequence, stronger surface currents, the equatorial upwelling in the Pacific and Atlantic intensified.
5. The thermal contrast between surface and bottom waters was less than it is today.
6. The sea level was lower by at least 85 m.

The CLIMAP project further attempted to reconstruct the Last Interglacial Ocean (a time of minimum ice volume ~122 ka BP). It was the last time when global ice volume was comparable to that of today. Based on microfaunal census counts and oxygen isotopic composition of foraminifera across the world's oceans, the following conclusions were made (CLIMAP Project Members 1984):

1. The last interglacial ocean was not significantly different from the modern oceans.
2. The sea-surface temperatures were little different from those today. There is some evidence of a slightly warmer North Atlantic and North Pacific and a cooler Gulf of Mexico.
3. The peak of warmth of the last interglaciation was regionally diachronous by several thousand years.

Gateways and Ocean Circulation

The topographic changes in the ocean basins, climate and plate tectonics shape the ocean water circulation pattern. Narrow passages formed due to the changing position of the continents may restrict the flow of ocean water. Such narrow conduits are termed ocean gateways and they have important influence on the distribution of surface water masses and vertical structure of the oceans. The Drake Passage connecting the Pacific and Atlantic Oceans and the Indonesian Seaway connecting the Indian and Pacific Oceans are examples of gateways. The shifting continents through the Cenozoic leading to opening of the gateways at high latitudes and their closing in the tropics (Fig. 13.6) had major influence on ocean circulation, climate and biotic changes.

Extensive paleobiogeographic studies carried out on deep-sea cores in the 1980s provided valuable clues about the importance of microfossils in ascertaining the precise timing of the changing plate boundary conditions linked to the opening and closing of major ocean gateways. For instance, the paleobiogeographic distribution of planktic foraminifer *Jenkinsella samwelli* provided the clue for the initiation of the Circum Antarctic Circulation during the Paleogene, caused by the opening of the Tasman Seaway and the Drake Passage. Likewise, the presence of *Pulleniatina spectabilis* (5.6–4.2 Ma) in the equatorial Pacific and its complete absence from both the Indian and Atlantic Oceans have been attributed to the blocking of tropical waters of the Pacific from going into the Indian and Atlantic Oceans due to the closing of the Indonesian seaway and the Panama Isthmus, respectively (Fig. 13.7). Similarly, there are at least three tropical planktic foraminiferal species (*Globorotaloides hexagona*, *Globoquadrina conglomerata*, *Globigerinella adamsi*) which occur in the Indo-Pacific region, but are absent from the Atlantic due to the

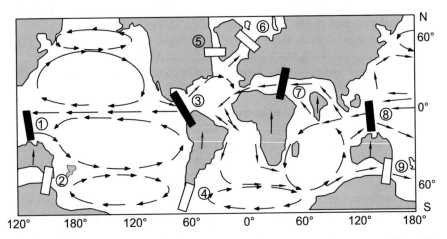

Fig. 13.6 The opening (*open rectangle*) and closing (*shaded rectangle*) of the gateways: (*1*) and (*8*) Indonesian Seaway, (*2*) Australian–Tasman–Antarctic Gateway, (*3*) Central American Seaway, (*4*) Drake Passage, (*5*) Iceland Norwegian Passage, (*6*) Denmark–Saeroe Passage, (*7*) Tethyan Seaway (redrawn after Seibold and Berger (1996), with permission © Springer Science + Business Media)

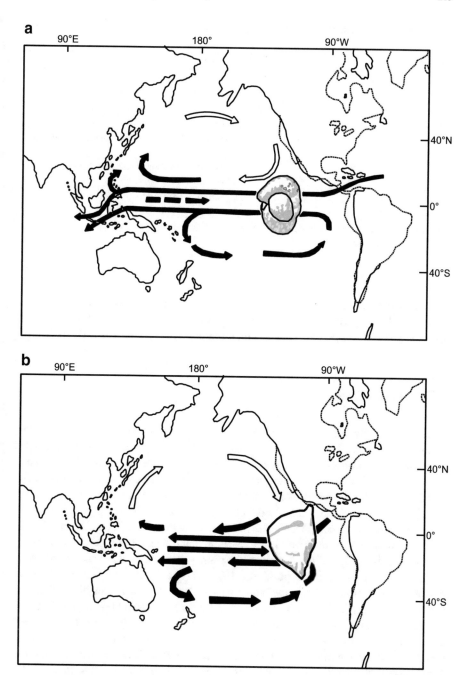

Fig. 13.7 Surface circulation in the Pacific during the Late Miocene (*top*) and early Pliocene (*bottom*), reconstructed based on biogeographic distribution of *Pulleniatina primalis* and *Pulleniatina spectabilis*. Note the change in flow due to closing of the Indonesian seaway in the early Pliocene (redrawn after Srinivasan and Sinha (1998), with permission ©Pergamon)

Table 13.3 Major Cenozoic Ocean Gateway events, and their timing and effect on ocean circulation

Gateway events	Time[a]	Effect on ocean circulation
Opening of Tasman Gateway	53–45 Ma[1]	Separation of Australia from Antarctica Separation of South America from Antarctica
Opening of Drake Passage	30–25 Ma[2]	*Development of Circum Antarctic Current and initiation of Psychrospheric Circulation*
Closing of Tethys Seaway	18 Ma[3]	Disconnection of eastern Mediterranean from Indian Ocean
Closing of Indonesian Seaway	5.6–4.2 Ma[4]	Blocking of equatorial current system between Pacific and Indian Oceans
Closing of Central American Seaway	3–2.6 Ma[5]	Disconnection of the tropical Atlantic from Pacific *Diminishing Equatorial circulation*

[a]Based on [1]Kennett (1977); [2]Kennett (1983); [3]Rogl and Steininger (1984); [4]Srinivasan and Sinha (2000); [5]Keigwin (1982)

closing of the Central American Seaway during the late Pliocene. The global synthesis of paleobiogeographic and other proxy data has provided clues to the opening and closing of the major gateways and their effect on ocean circulation (Table 13.3).

CCD Fluctuation

The calcite compensation depth (CCD) is the depth in the ocean where the rate of dissolution of carbonate balances the rate of accumulation. It is marked by a facies change from calcareous ooze to brown clay or siliceous ooze and has a mean depth of 4.5 km. The CCD in the Pacific is at shallower depths of 4200–4500 m, while in the Atlantic, it is at 5000 m or more. The state of preservation of microfossils is a proxy of dissolution of carbonates in the ocean water. The dissolution of calcareous tests of planktic foraminifera, pteropod and coccoliths starts at depths shallower than the CCD, and the depth that separates well-preserved assemblages of these microfauna from poorly preserved assemblages is the lysocline. Aragonite dissolves more readily than calcite and, as a result, the CCD for aragonite is positioned 1–2 km above the CCD of calcite. The CCD has fluctuated by nearly 2 km in the Mesozoic and Cenozoic (van Andel 1975). There is a general similarity in the trend of CCD across the major oceans (Fig. 13.8), indicating that the chemical environment of the oceans has changed globally. It was shallower in the Late Eocene, dropped markedly near the Eocene–Oligocene boundary, and became shallower again in the Miocene (10–15 million years before). The mechanism that produced marked fluctuations in the CCD has yet to be fully understood. The high stands correspond to shallower CCD and low stands to deeper CCD. The concept of basin-shelf fractionation explains the relationship between sea-level change and the rise and fall of CCD. During high stands, when a shelf is flooded, carbonate favourably accumulates on the shelf and the deep-sea floor starves due to removal of $CaCO_3$ from the ocean. During low stands, the shelf is bared and carbonate is supplied to the deep sea

Fig. 13.8 The fluctuation of Calcite Compensation Depth (CCD) in the Cenozoic oceans (redrawn after van Andel (1975), with permission ©Elsevier)

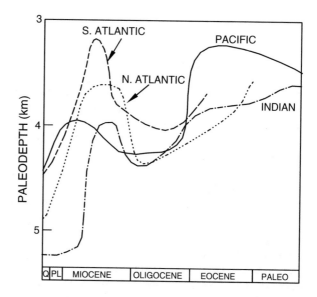

(Seibold and Berger 1996). The studies reveal that the pronounced carbonate minima (dissolution peaks) are closely associated with a marked increase in $\delta^{18}O$ of deep-sea benthic foraminifera, climate cooling and regression events.

Hiatus and Deep-Sea Circulation

High resolution biostratigraphic analysis of Cenozoic deep-sea cores has revealed a number of hiatuses of ocean-wide significance. Identification of these hiatuses has helped in studying the history and dynamics of bottom-water circulation. The hiatuses correspond to intervals of cold-water assemblages, rapid enrichment of $\delta^{18}O$ and sea-level fall (Barron and Keller 1982). Thus, the widespread Neogene deep-sea hiatuses recorded in the world's oceans testify to the development of strong bottom currents and corrosive deep waters reflecting periods of increased glaciations in polar regions and offer valuable clues about ancient deep circulation. In the Northern Indian Ocean, major deep-sea hiatuses are recorded at 23–22, 20–18, 16–15, 13.5–11 and 5.2–4.7 Ma, representing intervals of increased bottom-water activity and psychrospheric circulation (Srinivasan and Chaturvedi 1992; Singh and Srinivasan 1995).

The sluggish deep sea and well-circulated deep sea are also distinguished by aerobic and anaerobic foraminiferal assemblages, respectively. The oxygen index (Sect. 11.3) data compiled over the world's oceans (Kaiho 1991) suggest that low-oxygen deep-sea conditions developed in the oceans in the Early Eocene and Late Oligocene times. This corresponds to episodes of oceanic warming and sluggish deep-water circulation. The circulation was vigorous and the ocean bottom was oxygenated in the Late Maastrichtian–Middle Paleocene, Middle Eocene–Early Oligocene and Middle Miocene–Holocene cooling periods.

Paleoproductivity

Oceanic productivity is the uptake of dissolved inorganic carbon and its sequestration into organic compounds by marine primary producers. Only a part of the net primary production in the surface water reaches the deep ocean as an export production. The present-day net production in the open ocean is about 25–250 g C m^{-2} years^{-1} and export production is less than 10–50 % of this value. The organic carbon and phosphorus content of sediments could be the direct measures of productivity in past oceans, but it is important to note that organic carbon progressively gets oxidized while settling in the water column and only a fraction of the phosphorus is preserved in sediments.

The microfossil assemblages and microfossil-based geochemical proxies are widely applied in paleoproductivity reconstruction. The planktic foraminifera *Globigerina bulloides, Neogloboquadrina dutertrei* and *Globorotalia tumida* are indicative of high productivity at low latitudes. The benthic foraminifera *Uvigerina peregrina* and *Globobulimina* characterize high nutrient fluxes and *Cibicides wuellerstorfi* and *Cibicides cullenbergi* indicate low nutrient in the deep sea. The abundance of benthic foraminifera is also employed to reconstruct paleoproductivity. The rate of accumulation of benthic foraminifera may be correlated with the flux of organic carbon reaching the sea floor (and, therefore, the surface productivity). Based on the equation for paleoproductivity (Table 13.2), the productivity in the equatorial Pacific during the LGM is estimated to have been 1.5–2 times higher than the present productivity (Herguera and Berger 1991). Siliceous microfossil diatoms are also known to be good indicators of productivity of the overlying waters. They occur most conspicuously in areas of upwelling and highly productive marginal seas.

The geochemistry of microfossil shells is a widely used proxy of productivity. The nitrogen isotopic composition of organic matter, silicon isotopic composition of diatom frustules and Cd/Ca ratios of planktic foraminifera are some of the common indicators of nutrients (N, P, Fe, Si) in the ocean. In the process of photosynthesis by phytoplankton, the lighter isotope of carbon (^{12}C) is preferentially sequestered in the organic matter and, therefore, the residual inorganic pool gets enriched in heavier isotope (^{13}C). The settling organic matter in the deep sea, thus, returns ^{12}C to the deep-water inorganic pool. The difference in δ^{13}C planktic and deep-water benthic foraminifera is, therefore, a measure of productivity, and the larger the difference between the benthic and planktic δ^{13}C, the higher the productivity (Payton 2009). Some of the planktic foraminifera, *Globigerina bulloides*, for example, thrive in upwelling waters. It records a lower δ^{13}C signal of the subsurface water. The δ^{13}C of non-upwelling species would record higher δ^{13}C of the surface water. The difference in δ^{13}C between particular species ($\Delta \delta^{13}$C), thus, reflects productivity or intensity of upwelling. It is validated by a good correlation between $\Delta \delta^{13}$C and the marine organic carbon record in the Congo Fan (Schneider et al. 1994). The proxy employed to reconstruct the paleoproductivity of the Congo recorded a 23 ka periodicity of upwelling and productivity changes in the Quaternary.

References

Bandy OL (1960) The geologic significance of coiling ratios in the foraminifer *Globigerina pachyderma* (Ehrenberg). J Paleontol 34:671–681

Barron JA, Keller G (1982) Widespread Miocene deep sea hiatuses: coincidence with periods of global cooling. Geology 10:577–581

Berger WH (1969) Ecologic patterns of living planktonic foraminifera. Deep-Sea Res 16:1–24

Bharti SK, Singh AD (2013) *Bulimina arabiensis*, a new species of benthic foraminifera from the Arabian Sea. J Foraminiferal Res 43:255–261

CLIMAP Project Members (1976) The surface of the ice-age earth. Science 191(4232):1131–1137

CLIMAP Project Members (1984) The last interglacial ocean. Quat Res 21:123–224

Emiliani C (1955) Pleistocene temperatures. J Geol 63:538–578

Gooday AJ (2003) Benthic Foraminifera (Protista) as tools in deep-water palaeoceanography: environmental influences on faunal characteristics. In: Southward AJ, Tyler PA, Young CM, Fuiman LA (eds) Advances in marine biology, 46th edn. Amsterdam, Academic, pp 3–90

Guiot J, de Vernal A (2007) Transfer functions: methods for quantitative paleoceanography based on microfossils. In: Hillaire-Marcel C, de Vernal A (eds) Proxies in Late Cenozoic Paleoceanography. Elsevier, Amsterdam, pp 523–563

Gupta AK, Anderson DM, Overpeck JT (2003) Abrupt changes in the Asian southwest monsoon during the Holocene and their links to the North Atlantic Ocean. Nature 421:354–357

Herguera JC, Berger WA (1991) Paleoproductivity from benthic foraminifera abundance: glacial to postglacial change in the west-equatorial Pacific. Geology 19:1173–1176

Hillaire-Marcel C, de Vernal A (2007) Methods in late Cenozoic paleoceanography: introduction. In: Hillaire-Marcel C, de Vernal A (eds) Proxies in Late Cenozoic Paleoceanography. Elsevier, Amsterdam, pp 1–15

Imbrie J, Kipp NG (1971) A new micropaleontological method for quantitative paleoclimatology: application to a late Pleistocene Caribbean core. In: Turekian KK (ed) The Late Cenozoic glacial ages. Yale University Press, New Haven, pp 71–181

Jorissen FJ, Fontanier C, Thomas E (2007) Paleoceanographical proxies based on deep-sea benthic foraminiferal assemblage characteristics. In: Hillaire-Marcel C, de Vernal A (eds) Proxies in Late Cenozoic paleoceanography. Elsevier, Amsterdam, pp 263–326

Kaiho K (1991) Global changes of paleogene aerobic/anaerobic benthic foraminifera and deep sea circulation. Palaeogeogr Palaeoclimatol Palaeoecol 83:65–85

Kaiho K (1994) Benthic foraminiferal dissolved-oxygen index and dissolved-oxygen levels in the modern ocean. Geology 22:719–722

Keigwin LD Jr (1982) Isotopic paleoceanography of the Caribbean and East Pacific: role of Panama uplift in the late Neogene time. Science 217:350–353

Kennett JP (1977) Cenozoic evolution of Antarctic glaciations, the circum-Antarctic Ocean and their impact on global paleoceanography. J Geophys Res 82:3843–3860

Kennett JP (1983) Neogene paleoceanography and plankton evolution. Suid Afrikaanse Tydskrifvir Wetenskap 81:251–253

Meissner KJ, Montenegro A, Avis C (2009) Paleoceanography. In: Gornitz V (ed) Encyclopedia of paleoclimatology and ancient environments. Springer, The Netherlands, pp 690–695

Naidu PD, Malmgren BA (1996) A high-resolution record of late Quaternary upwelling along the Oman Margin, Arabian Sea based on planktonic foraminifera. Paleoceanography 11:129–140

Payton A (2009) Ocean paleoproductivity. In: Gornitz V (ed) Encyclopedia of paleoclimatology and ancient environments. Springer, The Netherlands, pp 643–651

Rogl F, Steininger FF (1984) Neogene paratethys, Mediterranean and Indo-Pacific seaways: implications for the paleobiogeography of marine and terrestrial biotas. In: Brenchley P (ed) Fossils and climate. John Wiley, New York, pp 171–200

Schneider RR, Muller PJ, Wefer G (1994) Late Quaternary paleoproductivity changes off the Congo deduced from stable carbon isotopes of planktonic foraminifera. Palaeogeogr Palaeoclimatol Palaeoecol 110:255–274

Seibold E, Berger WH (1996) The sea floor–an introduction to marine geology, IIIth edn. Springer, Heidelberg

Sen Gupta BK, Machain-Castillo ML (1993) Benthic foraminifera in oxygen-poor habitats. Mar Micropaleontol 20:183–201

Singh AD, Srinivasan MS (1995) Neogene planktic foraminiferal biochronology of the Central Indian Ocean, DSDP sites 237 and 238. J Geol Soc Ind 45:445–462

Skinner BJ, Murck BW (2011) The blue planet: an introduction to earth system science, 3rd edn. John Wiley, New York

Srinivasan MS, Chaturvedi SN (1992) Neogene planktonic foraminiferal biochronology of the DSDP sites along the Ninety east Ridge Northern Indian Ocean, Centenary of Japanese micropaleontology. Terra, Tokyo, pp 175–188

Srinivasan MS, Kennett JP (1975) Paleoceanographically controlled ultrastructural variation in Neogloboquadrina pachyderma (Ehrenberg) at DSDP site 284, South Pacific. Initial Reports Deep Sea Drilling Project, Washington, DC. 30: 709–721

Srinivasan MS, Sinha DK (1998) Early Pliocene closing of the Indonesian Seaway: evidence from northeast Indian Ocean and tropical Pacific deep sea cores. J Asian Earth Sci 16(1):29–44

Srinivasan MS, Sinha DK (2000) Ocean circulation in the tropical Indo-Pacific during early Pliocene (5.6-4.2 Ma): paleobiogeographic and isotopic evidence. Proc Ind Acad Sci (Earth Planet Sci) 109(3):315–328

Van Andel TH (1975) Mesozoic / Cenozoic calcite compensation depth and the global distribution of calcareous sediments. Earth Planet Sci Lett 26:187–194

Vincent E, Berger WH (1981) Planktonic foraminifera and their use in paleoceanography. In: Emiliani C (ed) The oceanic lithosphere, vol 7, The sea. John Wiley, New York, pp 1025–1119

Woodruff F (1985) Changes in Miocene benthic foraminiferal distribution in the Pacific Ocean: relationship to paleoceanography. Geol Soc Am 163:131–175

Further Reading

Hillaire-Marcel C, de Vernal A (eds) (2007) Proxies in Late Cenozoic paleoceanography. Elsevier, Amsterdam

Kennett JP (1982) Marine geology. Prentice-Hall, Upper Saddle River, NJ

Seibold E, Berger WH (1996) The sea floor – an introduction to marine geology, IIIth edn. Springer, Heidelberg

Index

© Springer International Publishing Switzerland 2016
P.K. Saraswati, M.S. Srinivasan, *Micropaleontology*,
DOI 10.1007/978-3-319-14574-7

Printed in the United States
By Bookmasters